Optical Engineering

Optical Engineering

Madelyn Crews

CLANRYE INTERNATIONAL
www.clanryeinternational.com

Clanrye International,
750 Third Avenue, 9th Floor,
New York, NY 10017, USA

ISBN: 978-1-64726-144-3

Cataloging-in-Publication Data

Optical engineering / Madelyn Crews.
p. cm.
Includes bibliographical references and index.
ISBN: 978-1-64726-144-3
1. Optical engineering. 2. Mechanical engineering. 3. Optics.
4. Optical instruments. I. Crews, Madelyn.
TA1520 .O68 2022
621.36--dc23

For information on all Clanrye International publications
visit our website at www.clanryeinternational.com

CLANRYE INTERNATIONAL

TABLE OF CONTENTS

PREFACE

This book aims to help a broader range of students by exploring a wide variety of significant topics related to this discipline. It will help students in achieving a higher level of understanding of the subject and excel in their respective fields. This book would not have been possible without the unwavered support of my senior professors who took out the time to provide me feedback and help me with the process. I would also like to thank my family for their patience and support.

The field of study which deals with the application of optics is referred to as optical engineering. It focuses on the design of optical instruments including lenses, telescopes, microscopes and other equipment that use the properties of light in physics and chemistry. Some of the other devices which use principles from optical engineering are optical sensors and measurement systems, laser, fiber optic communication systems and optical disc systems. Optical engineering metrology is a subfield of optical engineering which utilizes optical methods for the measurement of micro-vibrations with instruments like the laser speckle interferometer. This book explores all the important aspects of optical engineering in the present day scenario. Different approaches, evaluations, methodologies and studies related to this field have been included herein. This textbook is appropriate for students seeking detailed information in this area as well as for experts.

A brief overview of the book contents is provided below:

Chapter – Introduction

Optics is the branch of physics that studies the properties and behavior of light and its interaction with matter. Optical engineering deals with the designing of optical instruments like lenses, telescope, etc. Geometrical optics, physical optics, quantum optics, etc. fall under its domain. This is an introductory chapter which will briefly introduce all the significant aspects of optics and optical engineering.

Chapter – Optical Materials and Lens Design

The materials which are used to construct the devices that alter or control the electromagnetic radiation in the visible, ultraviolet, and infrared spectral regions are known as optical materials. Lens designing is the designing of lens and it focuses on optical lens design, opto-mechanical system design, and optical fabrication and testing. This chapter has been carefully written to provide an easy understanding of optical materials and lens design.

Chapter – Optical Instruments

Some of the optical instruments are telescope, microscope, binoculars, camera, optical interferometer, photometer, polarimeter, refractometer, optical spectrometer, optical mapping, lasers and autocollimator. These diverse types of optical instruments have been thoroughly discussed in this chapter.

Chapter – Optical Sensors

An optical sensor deals with the conversion of light rays into an electronic signal. The major types of optical sensors are fiber-optic sensor and electro-optical sensor. This chapter closely examines these optical sensors as well as their applications to provide an extensive understanding of the subject.

Chapter - Optical Communication

Optical communication uses optical fibers to carry signals at a distance via light. Some of its types are fiber-optic communication, optical wireless communication, free-space optical communication and visible light communication. The topics elaborated in this chapter will help in gaining a better understanding of these types of optical communication.

Chapter - Applications

Optical engineering has a wide-range of applications of optics. Its applications include optical sensors, measurement systems, lasers, fiber optic communication systems, optical disc systems, photonic devices, biomedical optics, etc. The diverse applications of optical engineering have been thoroughly discussed in this chapter.

Madelyn Crews

Introduction

CHAPTER 1

Optics is the branch of physics that studies the properties and behavior of light and its interaction with matter. Optical engineering deals with the designing of optical instruments like lenses, telescope, etc. Geometrical optics, physical optics, quantum optics, etc. fall under its domain. This is an introductory chapter which will briefly introduce all the significant aspects of optics and optical engineering.

OPTICS

Optics is a science concerned with Optics, science concerned with the genesis and propagation of light, the changes that it undergoes and produces, and other phenomena closely associated with it. There are two major branches of optics, physical and geometrical. Physical optics deals primarily with the nature and properties of light itself. Geometrical optics has to do with the principles that govern the image-forming properties of lenses, mirrors, and other devices that make use of light. It also includes optical data processing, which involves the manipulation of the information content of an image formed by coherent optical systems.

Originally, the term optics was used only in relation to the eye and vision. Later, as lenses and other devices for aiding vision began to be developed, these were naturally called optical instruments, and the meaning of the term optics eventually became broadened to cover any application of light, even though the ultimate receiver is not the eye but a physical detector, such as a photographic plate or a television camera. In the 20th century optical methods came to be applied extensively to regions of the electromagnetic radiation spectrum not visible to the eye, such as X-rays, ultraviolet, infrared, and microwave radio waves, and to this extent these regions are now often included in the general field of optics.

In the present article the image-forming properties of lenses, mirrors, and other devices that make use of light are considered. The wave and quantum nature of light, its velocity, wavelength, polarization, diffraction, and interference may be found in light. The analysis of light into its component colours by prisms and gratings forms the basis of the extensive field of spectroscopy, the principles of which are discussed in spectroscopy. For information about the reception of light by the retina of the eye and the interpretation of images by the brain.

Geometrical Optics

The Optical Image

An optical image may be regarded as the apparent reproduction of an object by a lens or mirror system, employing light as a carrier. An entire image is generally produced simultaneously, as by the lens in a camera, but images may also be generated sequentially by point-by-point scanning, as in a television system or in the radio transmission of pictures across long distances in space. Nevertheless, the final detector of all images is invariably the human eye, and, whatever means is used to transmit and control the light, the final image must either be produced simultaneously or scanned so rapidly that the observer's persistence of vision will give him the mental impression of a complete image covering a finite field of view. For this to be effective the image must be repeated (as in motion pictures) or scanned (as in television) at least 40 times a second to eliminate flicker or any appearance of intermittency.

To the ancients, the processes of image formation were full of mystery. Indeed, for a long time there was a great discussion as to whether, in vision, something moved from the object to the eye or whether something reached out from the eye to the object. By the beginning of the 17th century, however, it was known that rays of light travel in straight lines, and in 1604 Johannes Kepler, a German astronomer, published a book on optics in which he postulated that an extended object could be regarded as a multitude of separate points, each point emitting rays of light in all directions. Some of these rays would enter a lens, by which they would be bent around and made to converge to a point, the "image" of the object point whence the rays originated. The lens of the eye was not different from other lenses, and it formed an image of external objects on the retina, producing the sensation of vision.

There are two main types of image to be considered: real and virtual. A real image is formed outside the system, where the emerging rays actually cross; such an image can be caught ona screen or piece of film and is the kind of image formed by a slide projector or in a camera. A virtual image, on the other hand, is formed inside an instrument at the point where diverging rays would cross if they were extended backward into the instrument. Such an image is formed in a microscope or telescope and can be seen by looking into the eyepiece.

Kepler's concept of an image as being formed by the crossing of rays was limited in that it took no account of possible unsharpness caused by aberrations, diffraction, or even defocussing. In 1957 the Italian physicist Vasco Ronchi went the other way and defined an image as any recognizable nonuniformity in the light distribution over a surface such as a screen or film; the sharper the image, the greater the degree of nonuniformity. Today, the concept of an image often departs from Kepler's idea that an extended object can be regarded as innumerable separate points of light, and it is sometimes more convenient to regard an image as being composed of overlapping patterns of varying frequencies and contrasts; hence, the quality of a lens can be expressed by a graph connecting the spatial frequency of a parallel line object with the contrast in the image.

Optics had progressed rapidly by the early years of the 19th century. Lenses of moderately good quality were being made for telescopes and microscopes, and in 1841 the great mathematician Carl Friedrich Gauss published his classical book on geometrical optics. In it he expounded the concept of the focal length and cardinal points of a lens system and developed formulas for calculating the position and size of the image formed by a lens of given focal length. Between 1852 and 1856 Gauss's theory was extended to the calculation of the five principal aberrations of a lens, thus laying the foundation for the formal procedures of lens design that were used for the next 100 years. Since about 1960, however, lens design has been almost entirely computerized, and the old methods of designing lenses by hand on a desk calculator are rapidly disappearing.

By the end of the 19th century numerous other workers had entered the field of geometrical optics, notably an English physicist, Lord Rayleigh (John William Strutt), and a German physicist, Ernst Karl Abbe. It is impossible to list all their accomplishments here. Since 1940 there has been a great resurgence in optics on the basis of information and communication theory, which is treated at length below.

Light Rays, Waves and Wavelets

A single point of light, which may be a point in an extended object, emits light in the form of a continually expanding train of waves, spherical in shape and centred about the point of light. It is, however, often much more convenient to regard an object point as emitting fans of rays, the rays being straight lines everywhere perpendicular to the waves. When the light beam is refracted by a lens or reflected by a mirror, the curvature of the waves is changed, and the angular divergence of the ray bundle is similarly changed in such a way that the rays remain everywhere perpendicular to the waves. When aberrations are present, a convergent ray bundle does not shrink to a perfect point, and the emerging waves are then not truly spherical.

In 1690 Christiaan Huygens, a Dutch scientist, postulated that a light wave progresses because each point in it becomes the centre of a little wavelet travelling outward in all directions at the speed of light, each new wave being merely the envelope of all these expanding wavelets. When the wavelets reach the region outside the outermost rays of the light beam, they destroy each other by mutual interference wherever a crest of one wavelet falls upon a trough of another wavelet. Hence, in effect, no waves or wavelets are allowed to exist outside the geometrical light beam defined by the rays. The normal destruction of one wavelet by another, which serves to restrict the light energy to the region of the rectilinear ray paths, however, breaks down when the light beam strikes an opaque edge, for the edge then cuts off some of the interfering wavelets, allowing others to exist, which diverge slightly into the shadow area. This phenomenon is called diffraction, and it gives rise to a complicated fine structure at the edges of shadows and in optical images.

The Pinhole Camera

An excellent example of the working of the wavelet theory is found in the well-known pinhole camera. If the pinhole is large, the diverging geometrical pencil of rays leads to a blurred image, because each point in the object will be projected as a finite circular patch of light onthe film. The spreading of the light at the boundary of a large pinhole by diffraction is slight. If the pinhole is made extremely small, however, the geometrical patch then becomes small, but the diffraction spreading is now great, leading once more to a blurred picture. There are thus two opposing effects present, and at the optimum hole size the two effects are just equal. Thisoccurs when the hole diameter is equal to the square root of twice the wavelength (λ) times the distance (f) between the pinhole and film—i.e., Square root of $\sqrt{2\lambda}$ f. For f = 100 millimetres and λ = 0.0005 millimetre, the optimum hole size becomes 0.32 millimetre. This is not very exact, and a 0.4-millimetre hole would probably be just as good in practice. A pinhole, like a camera lens, can be regarded as having an f-number, which is the ratio of focal length to aperture. In this example, the f-number is 100/0.32 = 310, designated f/310. Modern camera lenses have much greater apertures, in order to achieve light-gathering power, of around f/1.2–f/5.6.

Resolution and the Airy Disk

When a well-corrected lens is used in place of a pinhole, the geometrical ray divergence is eliminated by the focussing action of the lens, and a much larger aperture may be employed; in that case the diffraction spreading becomes small indeed. The image of a point formed by a perfect lens is a minute pattern of concentric and progressively fainter rings of light surrounding a central dot, the whole structure being called the Airy disk after George Biddell Airy, an English astronomer, who first explained the phenomenon in 1834. The Airy disk of a practical lens is small, its diameter being approximately equal to the f-number of the lens expressed in microns (0.001 millimetre). The Airy disk of an f/4.5 lens is therfore about 0.0045 millimetre in diameter (ten times the wavelength of blue light). Nevertheless, the Airy disk formed by a telescope or microscope objective can be readily seen with a bright point source of light if a sufficiently high eyepiece magnification is used.

The finite size of the Airy disk sets an inevitable limit to the possible resolving power of a visual instrument. Rayleigh found that two adjacent and equally bright stars can just be resolved if the image of one star falls somewhere near the innermost dark ring in the Airy disk of the other star; the resolving power of a lens can therefore be regarded as about half the f-number of the lens expressed in microns. The angular resolution of a telescope is equal to the angle subtended by the least resolvable image separation at the focal length of the objective, the light-gathering lens. This works out at about four and a half seconds of arc divided by the diameter of the objective in inches.

The Rayleigh Limit

When a perfect lens forms an image of a point source of light, the emerging wave is a sphere centred about the image point. The optical paths from all points on the wave to the image are therefore equal, so that the expanding wavelets are all in phase (vibrating in unison) when they reach the image. In an imperfect lens, however, because of the presence of aberrations, the emerging wave is not a perfect sphere, and the optical paths from the wave to the image point are then not all equal. In such a case some wavelets will reach the image as a peak, some as a trough, and there will be much destructive interference leading to the formation of a sizable patch of light, much different from the minute Airy disk characteristic of a perfectly corrected lens. In 1879 Rayleigh studied the effects of phase inequalities in a star image and came to the conclusion that an image will not be seriously degraded unless the path differences between one part of the wave and another exceed one-quarter of the wavelength of light. As this difference represents only 0.125 micron (5×10^{-6} inch), it is evident that an optical system must be designed and constructed with almost superhuman care if it is to give the best possible definition.

Reflection

The use of polished mirrors for reflecting light has been known for thousands of years, and concave mirrors have long been used to form real images of distant objects. Indeed, Isaac Newton greatly preferred the use of a mirror as a telescope objective to the poor-quality lenses available in his time. Because there is no limit to the possible size of a mirror, all large telescopes today are of this type.

When a ray of light is reflected at a polished surface, the angle of reflection between ray and normal (the line at right angles to the surface) is exactly equal to the angle of incidence. It can be seen that a convex mirror forms a virtual image of a distant object, whereas a concave mirror forms a real image. A plane mirror forms a virtual image of near objects, as in the familiar looking glass. Plane mirrors are often used in instruments to bend a beam of light into a different direction.

The Law of Refraction

When a ray of light meets the surface of separation between two transparent media, it is sharply bent or refracted. Because rays are really only directions and have no physical existence, the passage of light waves through a surface must be considered if refraction is to be understood. Refraction effects are based on the fact that light travels more slowly in a denser medium. The ratio of the velocity of light in air to its velocity in the medium is called the refractive index of the medium for light of a particular colour or wavelength. The refractive index is higher for blue light than for light at the red end of the spectrum.

In figure, AA′ represents a plane wave of light at the instant that A′ meets the plane refracting surface A′B separating two media having refractive indices n and n′, respectively. During the time taken by the light to travel from A to B in material n, light travels from A′ to B′ in material of refractive index n′, forming the new wave BB′ in the second material, proceeding in direction BC. Hence, the relationship n′/n = AB/A′B′ can be obtained; and dividing numerator and denominator by BA′ gives:

$$\frac{n'}{n} = \frac{AB / BA'}{A'B' / BA'} = \frac{\sin I}{\sin I'}$$

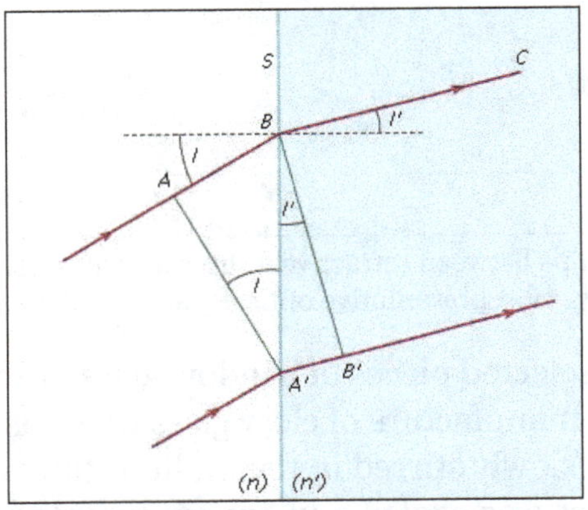

The law of refraction. Plane light wave at position AA′ in medium of index n and BB′ in medium of index n′.

The angles I and I′ are called the angle of incidence and angle of refraction between the refracting surface and the incident and refracted waves, respectively.

Returning now to the convention of considering the movement of light in terms of rays because entering and emerging rays are always perpendicular to the light waves they represent, angles I and I′ also denote the angles between the entering and emerging rays and the normal (perpendicular) line to the refracting surface at B.

Equation $\frac{n'}{n} = \frac{AB / BA'}{A'B' / BA'} = \frac{\sin I}{\sin I'}$, known as the law of refraction, is generally written: n′ sin I′ = n sin I.

Dispersion

The difference between the refractive indices of a transparent material for a specific blue light and a specific red light is known as the dispersion of the material. The usual choices of blue and red lights are the so-called “F” and “C” lines of hydrogen in the solar spectrum, named by Fraunhofer, with wavelengths 4861 and 6563 angstroms (the angstrom unit, abbreviated Å, is 10^{-8} centimetre), respectively. It is generally more significant, however, to compare the dispersion with the mean refractive index of the material for some intermediate colour such as the sodium “D” Fraunhofer line of wavelength 5893 angstroms. The dispersive power (w) of the material is then defined as the ratio of the difference between the “F” and “C” indices and the “D” index reduced by 1, or,

$$w = \frac{{}^{n}F - {}^{n}C}{{}^{n}D - 1}$$

Hundreds of different types of optical glass are currently available from manufacturers. These may be represented graphically on a plot of mean refractive index against dispersive power.

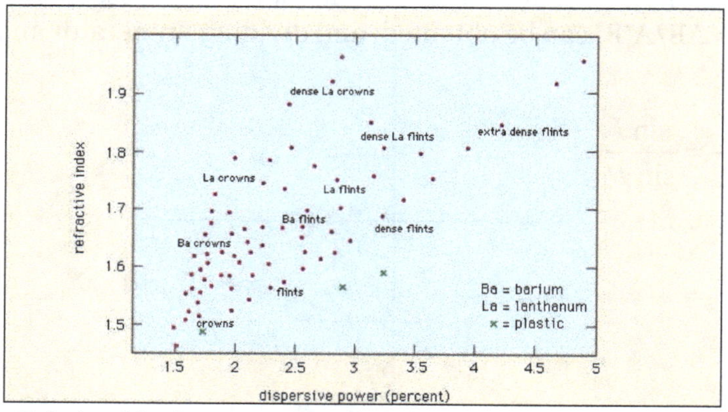

Relationships between refractive indices and dispersive powers of several representative optical glasses and plastics.

At first lenses were made from selected pieces of window glass or the glass used to make blown tableware. In the early 1800s, the manufacture of clear glass that was intended specifically for lenses began in Europe. The glass was slowly stirred in the molten state to remove striations and irregularities, and then the whole mass was cooled and broken up into suitable pieces for lens making. Subsequently, the pieces were placed in molds of the approximate size of the lens, slowly remelted to shape, and carefully annealed; i.e., allowed to cool slowly under controlled conditions to reduce strains and imperfections. Various chemicals were added in the molten state to vary the properties of the glass: addition of lead oxide, for example, was found to raise both the refractive index and the dispersive power. In 1884 it was discovered that barium oxide had the effect of raising the refractive index without increasing the dispersion, a property that proved to be of the greatest value in the design of photographic lenses known as anastigmats (lenses devoid of astigmatic aberration). In 1938 a further major improvement was achieved by the use of various rare-earth elements, and since 1950 lanthanum glass has been commonly used in high-quality photographic lenses.

The cost of optical glass varies considerably, depending on the type of glass, the precision with which the optical properties are maintained, the freedom from internal striae and strain, the number of bubbles, and the colour of the glass. Many common types of optical glass are now available in quite large pieces, but as the specifications of the glass become more stringent the cost rises and the range of available sizes becomes limited. In a small lens such as a microscope objective or a telescope eyepiece, the cost of the glass is insignificant, but in large lenses in which every millimetre of thickness may represent an additional pound in weight, the cost of the glass can be very high indeed.

Lenses can be molded successfully of various types of plastic material, polymethyl methacrylate being the most usual. Even multi-element plastic lenses have been manufactured for low-cost cameras, the negative (concave) elements being made of a high-dispersion plastic such as styrene.

Total Internal Reflection

When a ray of light emerges obliquely from glass into air, the angle of refraction between ray and normal is greater than the angle of incidence inside the glass, and at a sufficiently high obliquity the angle of refraction can actually reach 90°. In this case the emerging ray travels along the glass surface, and the sine of the angle of incidence inside the glass, known as the critical angle, is then equal to the reciprocal of the refractive index of the material. At angles of incidence greater than the critical angle, the ray never emerges, and total internal reflection occurs, for there is no measurable loss if the glass surface is perfectly clean. Dirt or dust on the surface can cause a small loss of energy by scattering some light into the air.

The hair-thin fibres used in fibre optics.

Light is totally internally reflected in many types of reflecting prism and in fibre optics, in which long fibres of high-index glass clad with a thin layer of lower index glass are assembled side-by-side in precise order. The light admitted into one end of each fibre is transmitted along it without loss by thousands of successive internal reflections at the interlayer between the glass and the cladding. Hence, an image projected upon one end of the bundle will be dissected and transmitted to the other end, where it can be examined through a magnifier or photographed. Many modern medical instruments, such as cystoscopes and bronchoscopes, depend for their action on this principle. Single thick fibres (actually glass rods) are sometimes used to transmit light around corners to an otherwise inaccessible location.

Ray-tracing Methods

Graphical Ray Tracing

In 1621 Willebrord Snell, a professor of mathematics at Leiden, discovered a simple graphical procedure for determining the direction of the refracted ray at a surface when the incident ray is given. The mathematical form of the law of refraction, equation $\frac{n'}{n} = \frac{AB / BA'}{A'B'/ BA'} = \frac{\sin I}{\sin I'}$ was announced by the French mathematician René Descartes some 16 years later.

Snell's construction is as follows: The line AP in figure below represents a ray incident upon a refracting surface at P, the normal at P being PN. If the incident and refracted rays are extended to intersect any line SS parallel to the normal, the lengths PQ and PR along the rays will be proportional to the refractive indices n and n′. Hence, if PQ and the indices are known, PR can be found and the refracted ray drawn in.

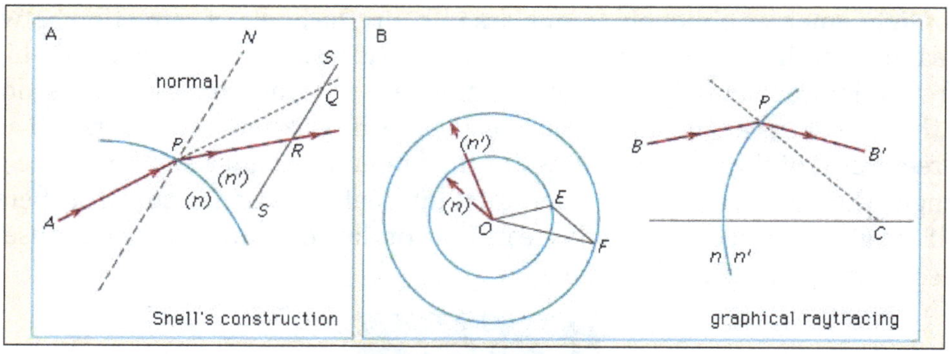

Graphic refraction procedures.

A convenient modification of Snell's construction can readily be used to trace the path of a ray through a complete lens. In figure, the incident ray BP strikes a refracting surface at P. The normal to the surface is PC. At any convenient place on the page two concentric circles are drawn about a point O with radii proportional to the refractive indices n and n′, respectively. A line OE is now drawn parallel to the incident ray BP extending as far as the circle representing the refractive index n of the medium containing the incident ray. From E a line is drawn parallel to the normal PC extending to F on the circle representing the refractive index n′. The line OF then represents the direction of the desired refracted ray, which may be drawn in at PB′. This process is repeated successively for all the surfaces in a lens. If a mirror is involved, the reflected ray may be found by drawing the normal line EF across the circle diagram to the incident-index circle on the other side.

Trigonometrical Ray Tracing

No graphical construction can possibly be adequate to determine the aberration residual of a corrected lens, and for this an accurate trigonometrical computation must be made and carried out to six or seven decimal places, the angles being determined to single seconds of arc or less. There are many procedures for calculating the path of a ray through a system of spherical refracting or reflecting surfaces, the following being typical: The diagram in Figure 4represents a ray lying in the meridian plane, defined as the plane containing the lens axis and the object point. A ray in this plane is defined by its slope angle, U, and by the length of the perpendicular, Q, drawn from the vertex (A) of the surface on to the ray. By drawing a line parallel to the incident ray through the centre of curvature C, to divide Q into two parts at N, the relation is stated as AN = r sin U, and NM = r sin I. Hence,

$$Q = r(\sin U + \sin I).$$

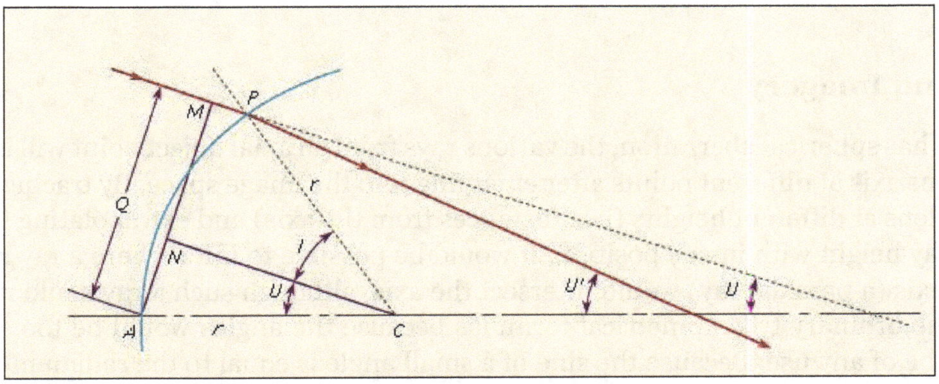

Trigonometrical ray tracing.

From this the first ray-tracing equation can be derived,

$$\sin I = \frac{Q}{r} - \sin U.$$

Applying the law of refraction, equation ($Q = r(\sin U + \sin I)$), gives the second equation,

$$\sin I' = \frac{n}{n'} = = -\sin I$$

Because the angle PCA = U + I = U′ + I′, the slope of the refracted ray can be written as,

$$U' = U + l - l'$$

and lastly, by adding primes to equation ($Q = r(\sin U + \sin I)$).

$$Q' = r(\sin U' + \sin I').$$

Having found the Q′ of the refracted ray, transfer to the next surface can be performed by,

$$Q_2 = Q_1' - d \sin U_3'.$$

in which d is the axial distance from the first to the second refracting surface. After performing this calculation for all the surfaces in succession, the longitudinal distance from the last surface to the intersection point of the emergent ray with the lens axis is found by,

$$L' = \frac{Q'}{\sin U'}.$$

Corresponding but much more complicated formulas are available for tracing a skew ray, that is, a ray that does not lie in the meridian plane but travels at an angle to it. After refraction at a surface, a skew ray intersects the meridian plane again at what is called the diapoint. By tracing the paths of a great many (100 or more) meridional and skew rays through a lens, with the help of an electronic computer, and plotting the assemblage of points at which all these rays pierce the focal plane after emerging from the lens, a close approximation to the appearance of a star image can be constructed, and a good idea of the expected performance of a lens can be obtained.

Paraxial and Imagery

In a lens that has spherical aberration, the various rays from an axial object point will in general intersect the lens axis at different points after emerging into the image space. By tracing several rays entering the lens at different heights (i.e., distances from the axis) and extrapolating from a graph connecting ray height with image position, it would be possible to infer where a ray running very close to the axis (a paraxial ray) would intersect the axis, although such a ray could not be traced directly by the ordinary trigonometrical formulas because the angles would be too small for the sine table to be of any use. Because the sine of a small angle is equal to the radian measure of the angle itself, however, a paraxial ray can be traced by reducing the ray-tracing formulas to their limiting case for small angles and thus determining the paraxial intersection point directly. When this is done, writing paraxial-ray data with lowercase letters, it is found that the Q and Q′ above both become equal to the height of incidence y, and the formulas $\sin I = \frac{Q}{r} - \sin U$, $\sin I' = \frac{n}{n'} - \sin I$, and $U' = U + l - l'$ become, in the paraxial limit:

$$i = \frac{y}{r'} - u$$

$$i' = \frac{n}{n'} i$$

$$u' = u + i - i'.$$

The longitudinal distance from the last surface to the intersection point of the emerging paraxial ray with the lens axis becomes $l' = y/u'$.

Because all paraxial rays from a given object point unite at the same image point, the resulting longitudinal distance (l′) is independent of the particular paraxial ray that is traced. Any nominal value for the height of incidence, y, may therefore be adopted, remembering that it is really an infinitesimal and y is only its relative magnitude. Thus, it is clear that the paraxial angles in below given equation are really only auxiliaries, and they can be readily eliminated, giving the object–image distances for paraxial rays:

$$i = \frac{y}{r'} - u$$

$$i' = \frac{n}{n'} i$$

$$u' = u + i - i'$$

$$n'(l' - r)u' = n(l - r)u \,.$$

$$\frac{n'}{l'} = \frac{n}{l} + \frac{n' - n}{r}.$$

Magnification: The Optical Invariant

It is frequently as important to determine the size of an image as it is to determine its location. To obtain an expression for the magnification—that is, the ratio of the size of an image to the size of the object—the following process may be used: If an object point *B* lies to one side of the lens axis at a transverse distance *h* from it, and the image point *B′* is at a transverse distance *h′*, then *B*, *B′*, and the centre of curvature of the surface, *C*, lie on a straight line called the auxiliary axis. Then, by simple proportion,

$$m = \frac{h'}{h} = \frac{l' - r}{l - r} = \frac{nu}{n'u'}.$$

Hence,

$$h'n'u' \quad hnu$$

and the product (*hnu*) is invariant for all the spaces between the lens surfaces, including the object and image spaces, for any lens system of any degree of complexity. This theorem has been named after the French scientist Joseph-Louis Lagrange, although it is sometimes called the Smith-Helmholtz theorem, after Robert Smith, an English scientist, and Hermann Helmholtz, a German scientist; the product (*hnu*) is often known as the optical invariant. As it is easy to determine the quantities *h*, *n*, and *u* for the original object, it is only necessary to calculate *u′* by tracing a paraxial ray in order to find the image height *h′* for any lens. If the lens is used in air, as most lenses are, the refractive indices are both unity, and the magnification becomes merely $m = u / u'$.

The Gauss Theory of Lenses

In 1841 Gauss published a now famous treatise on optics in which he demonstrated that, so far as paraxial rays are concerned, a lens of any degree of complexity can be replaced by two principal, or nodal, points and two focal points, the distances from the principal points to their respective focal points being the focal lengths of the lens, and furthermore, that the two focal lengths are equal to one another when the refractive indices of object and image spaces are equal, as when a lens is used in air.

The principal and focal points may be defined as follows: Figure shows a lens system of any construction, with a bundle of rays entering from the left in a direction parallel to the lens axis. After refraction by the lens each ray will cross the axis at some point, and the entering and emerging portions of each ray are then extended until they intersect at a point such as Q. The locus of all the points Q is a surface of revolution about the lens axis known as the equivalent refracting locus of the lens. The point where this locus crosses the axis is called the principal point, P_2, and the central portion of the locus in the neighbourhood of the axis, which is virtually a plane perpendicular to the axis, is called the principal plane. The point where the emerging paraxial ray crosses the axis is called the focal point F_2, the distance from P_2 to F_2 being the (posterior) focal length f'. A similar situation exists for a parallel beam of light entering from the right, giving the anterior principal point P_1, the anterior focal point F_1, and the front focal length f. For a lens in air it can be shown that the two focal lengths are equal in magnitude but opposite in direction—i.e., if F_2 is to the right of P_2, then F_1 must lie to the left of P_1, as in the case of an ordinary positive lens (one that gives a real image). In a negative lens (one that gives a virtual image), F_2 lies to the left of P_2, and the posterior focal length f' is negative.

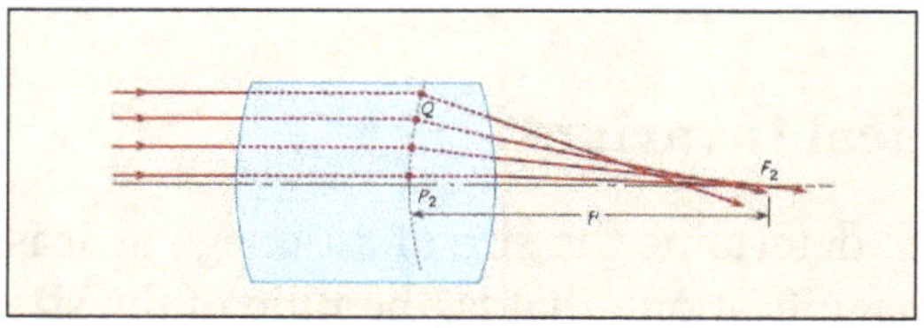

The Gauss theory.

The relation between the distances of object and image from a lens can be easily stated if the positions of the two principal points and the two focal points are known. (In using these expressions, distances are considered positive or negative depending on whether they are measured to the right or to the left from their respective origins.) For a lens in air: (a) If the conjugate distances measured from the respective focal points are x and x', and if m is the image magnification (height of image divided by height of object), then $m = -x'/f' = f'/x$ and $xx' = -f'^2$, (b) If the conjugate distances measured from the respective principal points are p and p' and if m is the image magnification, then $m = p'/p$ and $1/p' = 1/p + 1/f'$. The Lagrange equation ($h'n'u' = hnu$) requires modification for a distant object because in that case the object height h is infinite, and the slope angle u is zero. If the off-axis distance h is divided by the object distance L, and u is multiplied by L, equation ($h'n'u' = hnu$) becomes $h' = (n/n')f'\phi$, in which ϕ is the angle in radians subtended by the distant object at the lens. This formula provides a means for defining focal length and for measuring the focal length of an unknown lens.

The Thin Lens

In a thin lens such as a spectacle, the two principal planes coincide within the lens, and then the conjugate distances p and p' in the formula above become the distances of object and image from the lens itself.

The focal length of a thin lens can be computed by applying the surface-conjugate formula $h'n'u' = hnu$ to the two surfaces in succession, writing the l of the first surface as infinity and the l of the second surface equal to the l' of the first surface. When this is done, the lens power (P) becomes,

$$\frac{n'}{l'} = \frac{n}{l} + \frac{n'-n}{r}.$$

$$P = \frac{1}{f} = (n-1)\left(\frac{1}{r_1} - \frac{1}{r_2}\right).$$

Chromatic Aberration

Because the refractive index of glass varies with wavelength, every property of a lens that depends on its refractive index also varies with wavelength, including the focal length, the image distance, and the image magnification. The change of image distance with wavelength is known as chromatic aberration, and the variation of magnification with wavelength is known as chromatic difference of magnification, or lateral colour. Chromatic aberration can be eliminated by combining a strong lens of low-dispersion glass (crown) with a weaker lens made of high-dispersion (flint) glass. Such a combination is said to be achromatic. This method of removing chromatic aberration was discovered in 1729 by Chester Hall, an English inventor, and it was exploited vigorously in the late 18th century in numerous small telescopes. Chromatic variation of magnification can be eliminated by achromatizing all the components of a system or by making the system symmetrical about a central diaphragm. Both chromatic aberration and lateral colour are corrected in every high-grade optical system.

Longitudinal Magnification

If an object is moved through a short distance δp along the axis, then the corresponding image shift $\delta p'$ is related to the object movement by the longitudinal magnification (m). Succinctly,

$$\bar{m} = \delta p' / \delta p = m^2$$

in which m is the lateral magnification. The fact that the longitudinal magnification is equal to the square of the transverse magnification means that m is always positive; hence, if the object is moved from left to right, the image must also move from left to right. Also, if m is large, then m is very large, which explains why the depth of field (δp) of a microscope is extremely small. On the other hand, if m is small, less than one as in a camera, then m is very small, and all objects within a considerable range of distances (δp) appear substantially in focus.

Image of a Tilted Plane

If a lens is used to form an image of a plane object that is tilted relative to the lens axis, then the image will also be tilted in such a way that the plane of the object, the plane of the image, and the median plane of the lens all meet. This construction can be derived by the use of the lateral and longitudinal magnification relations just established above. With a tilted object the magnification at any point is given by the ratio of the distances of image and object from the lens at that point in the image, and, consequently, m varies progressively from one end of the image to the other. This arrangement is frequently used in view cameras equipped with "swings" to increase depth of field and in enlargers to rectify the convergence of parallel lines caused by tilting the camera, for example, in photographing tall buildings. The rule finds extensive application in photogrammetry and in the making of maps from aerial photographs.

Optical Systems

System Components

An optical system consists of a succession of elements, which may include lenses, mirrors, light sources, detectors, projection screens, reflecting prisms, dispersing devices, filters and thin films, and fibre-optics bundles.

Lenses

All optical systems have an aperture stop somewhere in the system to limit the diameter of the beams of light passing through the system from an object point. By analogy with the human eye, this limiting aperture stop is called the iris of the system, its images in the object and image spaces being called the entrance pupil and exit pupil, respectively. In most photographic lenses the iris is inside the objective, and it is often adjustable in diameter to control the image illumination and the depth of field. In telescope and microscope systems the cylindrical mount of the objective lens is generally the limiting aperture or iris of the system; its image, formed behind the eyepiece where the observer's eye must be located to see the whole area being observed, called the field, is then the exit pupil.

The pupils of a lens system can be regarded as the common bases of oblique beams passing through the system from all points in an extended object. In most systems, however, the mounts of some of the lens elements cut into the oblique beams and prevent the beams from being perfectly circular, and the pupils are then not fully filled with light. This effect is known as vignetting and leads to a reduction in illumination in the outer parts of the field of view.

A common feature of many optical systems is a relay lens, which may be introduced to invert an image or to extend the length of the system, as in a military periscope. An example of the use of a relay lens is found in the common rifle sight shown diagrammatically in figure below. Here the front lens *A* is the objective, forming an inverted image of the target on the cross wire or reticle at *B*. The light then proceeds to the relay lens *C*, which forms a second image, now erect, at *D*. Beyond this image is the eyepiece *E* to render the light parallel so that the image may be seen sharply by the observer. Unfortunately, the oblique beam from the objective will usually miss the relay lens, and so a field lens must be inserted at or near the first image *B* to bend the oblique beams around and redirect them toward the relay lens. The power of the field lens is chosen so that it will form an image of the objective lens aperture on the relay lens aperture. The iris and entrance pupil of this system coincide at the objective; there is an internal pupil at the relay lens, and the exit pupil lies beyond the eyepiece as shown in figure.

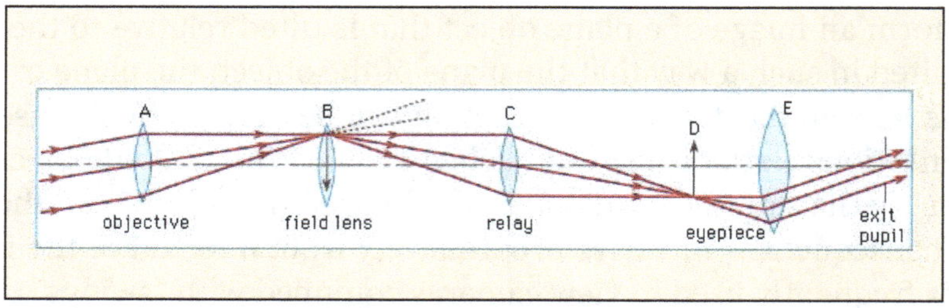

Operating principle of the telescopic rifle sight.

Mirrors

Mirrors are frequently used in optical systems. Plane mirrors may be employed to bend a beam of light in another direction, either for convenience or to yield an image reversed left for right if required. Curved mirrors, concave and convex, may be used in place of lenses as image-forming elements in reflecting telescopes. All of the world's largest telescopes and many small ones are of the reflecting type. Such telescopes use a concave mirror to produce the main image, a small secondary mirror often being added to magnify the image and to place it in a convenient position for observation or photography. Telescope mirrors are commonly made parabolic or hyperbolic in section to correct the aberrations of the image. Originally telescope mirrors were made from polished "speculum metal", an alloy of copper and tin, but in 1856 Justus von Liebig, a German chemist, invented a process for forming a mirror-like layer of silver on polished glass, which was applied to telescope mirrors by the German astronomer C.A. von Steinheil. Today most mirrors are made of glass, coated with either a chemically deposited silver layer or more often one made by depositing vaporized aluminum on the surface. The aluminum surface is as highly reflective as silver and does not tarnish as readily.

A large astronomical mirror presents many problems to the optical engineer, mainly because even a distortion of a few microns of the mirror under its own weight will cause an intolerable blurring of the image. Though many schemes for supporting a mirror without strain have been tried, including one to support it on a bag of compressed air, the problem of completely eliminating mirror distortion remains unsolved. A metal mirror, if well ribbed on the back, may be lighter than a glass mirror and therefore easier to handle, but most metals are slightly flexible and require just as careful support as glass mirrors. Since temperature changes can also cause serious distortion in a mirror, astronomers try to hold observatory temperatures as constant as possible.

Light Sources

Many types of optical instruments form images by natural light, but some, such as microscopes and projectors, require a source of artificial light. Tungsten filament lamps are the most common, but if a very bright source is required, a carbon or xenon arc is employed. For some applications, mercury or other gas discharge tubes are used; a laser beam is often employed in scientific applications. Laser light is brilliant, monochromatic, collimated (the rays are parallel), and coherent (the waves are all in step with each other), any or all of these properties being of value in particular cases.

Detectors

The image formed by an optical system is usually received by the eye, which is a remarkably adaptable and sensitive detector of radiation within the visible region of the electromagnetic spectrum. A photographic film, another widely used detector, has the advantage of yielding a permanent record of events. Since about 1925 many types of electrical detectors of radiation, both within the visible region and beyond it, have been developed. These include photoelectric cells of various kinds in which either a voltage or a resistance is modified by light falling on the device. Many new types of detectors are sensitive far into the infrared spectrum and are used to detect the heat radiated by a flame or other hot object. A number of image intensifiers and converters, particularly for X-ray or infrared radiation, which have appeared since World War II, embody a radiation detector at one end of a vacuum tube and an electron lens inside the tube to relay the image on to a phosphor screen at the other end. This arrangement produces a visible picture that may be observed by eye or photographed to make a permanent record.

Television camera tubes detect real images by electronic scanning, the picture on the viewing tube being a replica of the image in the original camera. The combined application of electronics and optics has become common. An extreme example of electro-optics appears in some space cameras, in which the film is exposed, processed, and then scanned by a tiny point of light; the light passing through the film is picked up by a photocell and transmitted to Earth by radio, where it is made to control the brightness of another point of light scanning a second piece of film in exact synchronism with the scanning spot in the camera. The whole system thus produces a picture on Earth that is an exact replica of the picture photographed in space a few minutes earlier.

Projection Screens

The simplest screen for the projection of slides or motion pictures is, of course, a matte white surface, which may be on a hard base as in outdoor theatres or on a stretched cloth indoors. A theatre screen is often perforated to transmit sound from loudspeakers placed behind it.

Improved screen materials have been developed to increase the brightness of the picture to suit the particular shape of the auditorium. A screen covered with tiny beads tends to send the light back in the general direction of the projector, and is suitable for use at one end of a long, narrow auditorium. Another type of screen is covered with fine embossed vertical grooves; this tends to distribute the light in a horizontal band across the audience with little or no vertical spread. A real advantage of these highly reflective screens is that they tend to reflect ambient room light away from the viewer as by a mirror, so that the pictures appear almost as bright and clear by day as in a darkened room.

Reflecting Prisms

Reflecting prisms are pieces of glass bounded by plane surfaces set at carefully specified angles. Some of these surfaces transmit light, some reflect light, while some serve both functions in succession. A prism is thus an assembly of plane reflectors at relatively fixed angles, which are traversed in succession by a beam of light.

The simplest prism is a triangular block of glass with two faces at right angles and one at an angle of 45°. The face at 45° deflects a beam of light through a right angle. The common Porro prism used in a pair of binoculars contains four 45° reflecting surfaces, two to reverse the beam direction in the vertical plane and two in the horizontal plane. These reflecting faces could be replaced by pieces of mirror mounted on a metal frame, but it is hard to hold mirrors rigidly and harder still to keep them clean. Some microscopes are equipped with a 45° deflection prism behind the eyepiece; this prism may provide two or three reflections depending on the type of image inversion or left-for-right reversal required.

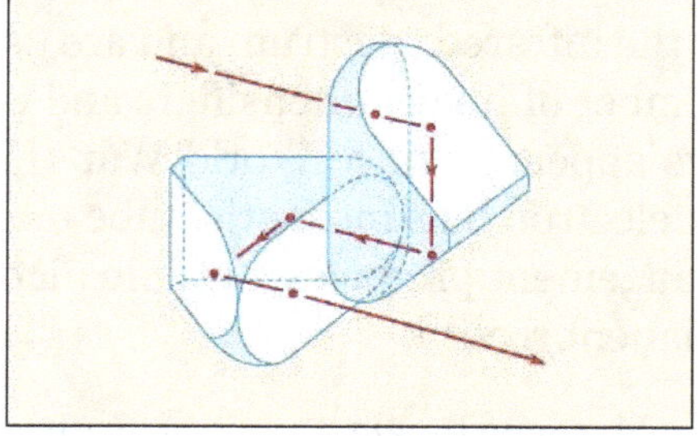

Porro prism.

Prisms containing a semireflecting, semitransmitting surface are known as beam splitters and as such have many uses. An important application is found in some colour television cameras, in which the light from the lens is divided by two beam splitters in succession to form red, green, and blue images on the faces of three image tubes in the camera.

Dispersing Devices

There are two forms of dispersing element used to spread out the constituent colours of a beam of light into a "spectrum," namely a prism and a grating. The prism, known to Newton, is the older; it separates the colours of the spectrum because the refractive index of the glass is lowest for red light and progressively increases through the yellow and green to the blue, where it is highest. Prism spectroscopes and spectrographs are made in a variety of forms and sizes, but in all cases the blue end of the spectrum is greatly spread out while the red end is relatively compressed.

A diffraction grating is a ruled mirror or transparent plate of glass having many thousands of fine parallel grooves to the inch. It separates the colours of the spectrum by a process of diffraction. Each groove diffracts, or scatters, light in all directions, and in the case of light of one particular wavelength, there will be one direction in which the light wave from one groove lags behind the light wave from the next groove by precisely one or more whole wavelengths. This results in a strong beam of diffracted light in that direction and darkness in all other directions. Since each spectral colour corresponds to a different wavelength, the grating spreads out the spectrum into a fan where it can be observed or photographed. The red rays are bent most and the blue rays least, the opposite of the situation with a prism.

Although a prism or grating is the essential dispersing element in a spectrograph, a fine slit and additional lenses or focussing mirrors must be used to form a sharply defined spectrum. Prism spectroscopes are, of course, limited to those wavelengths for which the prism material is transparent; a reflecting grating can be used for any wavelength that the material will reflect.

Filters and Thin Films

A colour filter is a sheet of transparent material that modifies a light beam by selective absorption of some colours in relation to others. A neutral filter absorbs all wavelengths equally and merely serves to reduce the intensity of a beam of light without changing its colour.

Filters may be made from sheets of coloured glass, plastic, or dyed gelatin, and in some cases glass cells filled with liquid have been used. Since World War II, another type of filterdepending on the interference of light has been developed in which one or more metallic or other types of films of controlled thickness have been deposited on a glass plate, the layers being so thin as to cause selective interference of some wavelengths in relation to others and thus act as a nonabsorbing filter. In this case the rejected colours are reflected instead of being absorbed.

Polarizing filters have the property of transmitting light that vibrates in one direction while absorbing light that vibrates in a perpendicular direction. These filters are used extensively in scientific instruments. In sunglasses and when placed over a camera lens, polarizing filters reduce unwanted reflections from nonmetallic surfaces. Polarizing spectacles have been used to separate the left-eye and right-eye beams in the projection of stereoscopic pictures or movies.

Fibre-optics Bundles

A thin rod or fibre of glass or other transparent material transmits light by repeated internal reflections, even when the rod is somewhat curved. An ordered bundle of rods or fibres is thus capable of taking an image projected upon one end of the bundle and reproducing it at the other end. A fibre-optics bundle can be fused together into a rigid channel, or it may be left flexible, only the ends being rigidly fastened together. Because a fibre bundle is exceedingly delicate, it must be handled with care; breaking a fibre would cause a black dot to appear in the reproduced image.

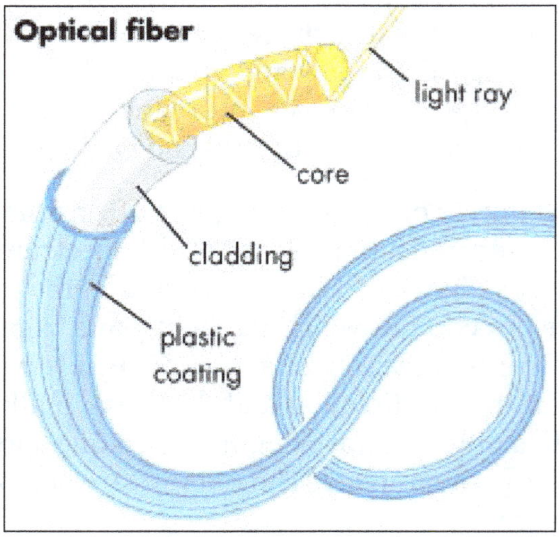

Light ray passing through an optical fibre.

Nonclassical Imaging Systems

Besides the familiar optical systems cited above, there are many nonclassical optical elements that are used to a limited extent for special purposes. The most familiar of these is the aspheric(non-spherical) surface. Because plane and spherical surfaces are the easiest to generate accurately on glass, most lenses contain only such surfaces. It is occasionally necessary, however, to use some other axially symmetric surface on a lens or mirror, generally to correct a particular aberration. An example is the parabolic surface used for the primary mirror of a large astronomical telescope; another is the elliptic surface molded on the front of the little solid glass reflector units used on highway signs.

Another commonly used optical surface is the side of a cylinder. Such surfaces have power only in the meridian perpendicular to the cylinder axis. Cylindrical lenses are therefore used wherever it is desired to vary the magnification from one meridian to a perpendicular meridian. Cylindrical surfaces are employed in the anamorphic lenses used in some wide-screen motion-picture systems to compress the image horizontally in the camera and stretch it back to its original shape in the projected image.

To correct astigmatism in the eye, many spectacles are made with toric surfaces—i.e., with a stronger curvature in one meridian than in the perpendicular meridian, like the bowl of a teaspoon. These surfaces are generated and polished by special machines and are made by the million every year.

Another nonclassical optical system is the bifocal or trifocal spectacle lens. They are made either by forming two or three separate surfaces on a single piece of glass or obtaining additional power by fusing a piece of high-index glass on to the front of the main lens and then polishing a single spherical surface over both glasses.

Two French scientists, Georges-Louis Buffon and Augustin-Jean Fresnel, in the 18th century suggested forming a lens in concentric rings to save weight, each ring being a portion of what would normally be a continuous spherical surface but flattened out. On a large scale, Fresnel lenses have been used in lighthouses, floodlights, and traffic signals, and as cylindrical ship's lanterns. With fine steps a few thousandths of an inch wide, molded plastic Fresnel lenses are often used as condensers in overhead projectors and in cameras as a field lens in contact with a ground-glass viewing screen.

Lenses have occasionally been made with one surface taking the form of a flattened cone. Such lenses produce a long, linear image of a point source, lying along the lens axis; for this reason they are commonly referred to as axicons. They have been used to produce a straight line of light in space for aligning machines and shafting, but since about 1965 the beam from a gas laser has generally been used instead.

Lens Aberrations

Seidel Sums

If a lens were perfect and the object were a single point of monochromatic light, then, as noted above, the light wave emerging from the lens would be a portion of a sphere centred about the ideal image point, lying in the paraxial image plane at a height above the axis given by the Lagrange theorem. In practice, however, this condition is most unlikely to occur; it is much more probable that the emerging wave will depart slightly from a perfect sphere, the departure varying from point to point over the lens aperture. This departure is extremely small, being of the order of the wavelength of light that is only half a micron, so it would be impossible to show this departure on a drawing. It can be represented mathematically, however, in the following way: The coordinates of a point in the exit-pupil aperture will be represented by x_0 and y_0, the y_0 coordinate lying in the meridian plane containing the object point and the lens axis. The departure of the wave from the ideal sphere is generally called OPD, meaning optical path difference. It can be shown that OPD is related to x_0 and y_0 by five constants S_1 through S_5, and the quantity h'_0,

$$\text{OPD} = S_1\,(x_0^2 + y_0^2)^2 + S_2 y_0 (x_0^2 + y_0^2) h_0' + S_3 (x_0^2 + 3y_0^2) h_0'^2 + S_4 (x_0^2 + y_0^2) h_0'^2 + S_5 y_0 h_0'^3$$

Each of these five terms is considered to be a separate "aberration," the coefficients S_1,...S_5, being called Seidel sums after the 19th-century German scientist L.P. Seidel, who identified the imperfections. These aberrations are respectively spherical, coma, astigmatism, Petzval field curvature, and distortion. The symbol h'_0 refers to the height of the final image point above the lens axis, and hence it defines the obliquity of the beam.

The five Seidel sums can be calculated by tracing a paraxial ray from object to image through the lens and by tracing also a paraxial principal ray from the centre of the aperture stop outward in both directions toward the object and image, respectively. The angle of incidence $i$and the ray slope angle u of each of these paraxial rays at each surface are then listed and inserted into the following expressions for the five sums. The angle u'_0 represents the final emerging slope of the paraxial ray.

The calculation starts by determining the radius A of the exit pupil by $A = \sqrt{x_o^2 + y_o^2}$ and also the quantity K at each surface by,

$$K = \frac{1}{2} yn\left(\frac{n}{n'} - 1\right)(i - u').$$

The corresponding K_{pr} for the paraxial principal ray is also determined at each surface. Then, the five aberrations may be written,

$$S_1 = \frac{1}{4A_4}\sum Ki^2, \qquad S_2 = \frac{1}{A^3 h_0'}\sum Kii_{pr}$$

$$S_3 = \frac{1}{2A^2 h_0'^2}\sum Ki_{pr}^2, \quad S_4 = \frac{u_0'^2}{4A^2}\sum \frac{n - n'}{nn'r}$$

$$S_5 = \frac{1}{Ah_0'^3}\sum\left[k_{pr} ii_{pr} + \frac{1}{2} h_0' u_0' (u_{pr}'^2 - u_{pr}^2)\right].$$

To interpret these aberrations, the simplest procedure is to find the components x', y' of the displacement of a ray from the Lagrangian image point in the paraxial focal plane, by differentiating the OPD expression given above. The partial derivatives $\partial OPD / \partial x_0$ and $\partial OPD / \partial y_0$ represent respectively the components of the slope of the wave relative to the reference sphere at any particular point (x_0, y_0). Hence, because a ray is always perpendicular to the wave, the ray displacements in the focal plane can be found by,

$$x' = f\frac{\partial OPD}{\partial x_0} \text{ and } y' = f\frac{\partial OPD}{\partial y_0}$$

in which f is the focal length of the lens. The aggregation of rays striking the focal plane will indicate the kind of image that is characteristic of each aberration.

This procedure will be applied to each of the five aberration terms separately, assuming that all the other aberrations are absent. Obviously, in a perfect lens x' and y' are zero because OPD is zero. It must be remembered, however, that by using rays instead of waves, all fine-structure effects caused by diffraction will be lost, and only the macroscopic image structure will be retained.

Spherical Aberration

The first term in the OPD expression is $\mathrm{OPD} = S_1\left(x_0^2 + y_0^2\right)^2$. Hence,

$$x' = f\frac{\partial OPD}{\partial x_0} = 4\mathrm{f}\, A^2 S_1 \cdot x_0 \text{ and } y' = f\frac{\partial OPD}{\partial y_0} = 4\mathrm{f}\, A^2 S_1 \cdot y_0$$

These displacements can both be eliminated simultaneously by applying a longitudinal shift L to the focal plane. This changes x' *by* $-Lx_0 / f$ and y' *by* $-Ly_0 / f$; hence, if L is made equal to $4f^2 A^2 S_1$, both ray displacements vanish. The aberration, therefore, represents a condition in which each zone of the lens has a different focus along the axis, the shift of focus from the paraxial image being proportional to A^2. This is known as spherical aberration.

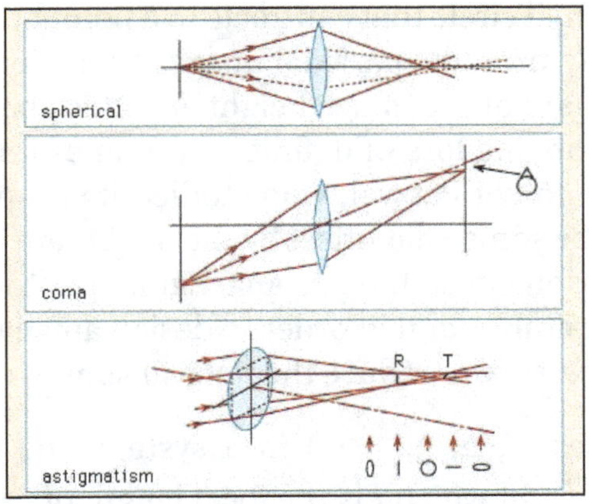

Lens aberrations.

Coma

The S_2 term in the OPD expression represents the aberration called coma, in which the image of a point has the appearance of a comet. The x' and y' components are as follows:

$$x' = fh_0' S_2 (2x_0 y_0)$$
$$y' = fh_0' S_2 (x_0^2 + 3y_0^2)$$

When this aberration is present, each circular zone of the lens forms a small ringlike image in the focal plane, the rings formed by successive concentric zones of the lens fitting into two straight envelope lines at 60° to each other. Because the brightness of this image is greatest at the tip, coma tends to form a one-sided haze on images in the outer parts of the field.

Astigmatism

If only the S_3 term is present, then:

$$x' = 2fh_0' S_3 (x_0)$$
$$y' = 2fh_0' S_3 (3y_0)$$

For any one zone of the lens, x' and y' describe a vertical ellipse with major axis three times the minor axis. The images formed by all the smaller zones of the lens fit into this ellipse and fill it out with a uniform intensity of light. If the image plane is moved along the axis by a distance L, as in focussing a camera, then, at $L = 2f^2 h_0'^2 S_3$, the ellipse shrinks to a radial focal line (R). Twice this displacement yields a circle; three times this L gives a tangential focal line (T), which is followed by an ellipse with its major axis in the x direction. The usual effect of astigmatism in an image is the appearance of radial or tangential blurring in the outer parts of the field.

Petzval Curvature

For the S_4 term taken alone:

$$x' = 2fh_0'^2 S_4 . x_0$$
$$y' = 2fh_0' S_4 \cdot y_0$$

The image of a point is now a small circle that contracts to a point at a new focus situated at a longitudinal distance $L = 2f^2 h_0'^2 S_4$ from the paraxial image. As the longitudinal displacement of the focus is proportional to the square of the image height h_0', this aberration represents a pure field curvature without any accompanying loss of definition (all lines remain sharp). It is named after the Hungarian mathematician József Petzval, who studied its properties in the early 1840s. The effect of Petzval curvature can be somewhat offset by the deliberate introduction of sufficient overcorrected astigmatism, as was done in all the pre-anastigmat photographic objectives. This added astigmatism is, of course, undesirable, and in order to design an anastigmat lens having a flat field free from astigmatism, it is necessary to reduce the Petzval sum S_4 drastically.

For a succession of thin lenses (1, 2, 3,...etc.) in a system, the Petzval sum becomes simply $1/f_1n_1 + 1/f_2n_2 + 1/f_3n_3 + \ldots$ etc., in which f is the focal length of each element and n is its refractive index. Therefore, to reduce the sum and minimize this aberration, relatively strong negative elements of low-index glass can be combined with positive elements of high-index glass. The positive and negative elements must be axially separated to provide the lens with a useful amount of positive power. The introduction of high-index barium crown glass with a low dispersive power in the 1880s initiated the development of anastigmat lenses.

Distortion

For the S_5 aberration,

$$x' = 0$$

$$y' = fh_0'^3 S_5.$$

When this aberration is present, the entire image point is displaced toward or away from the axis by an amount proportional to the third power of the transverse distance h_0' of the image from the axis. This leads to the formation of an image of a square that is either a barrel-shaped or a cushion-shaped figure.

It is to be noted that the five Seidel aberrations represent the largest and most conspicuous defects that can arise in an uncorrected optical system. Even in the best lenses in which these five aberrations have been perfectly corrected for one zone of the lens and for one point in the field, however, there will exist small residuals of these aberrations and of many other higher order aberrations also, which are significantly different from the classical types just described. The typical aberration shown in figure are, of course, grossly exaggerated, and actually it requires some magnification of a star image to render these appearances clearly visible. Nevertheless, they are important enough to require drastic reduction in high-quality lenses intended to make sharp negatives capable of considerable enlargement.

Image Brightness

General Relations

All photometric concepts are based on the idea of a standard candle, lamps having accurately known candle power being obtainable from the various national standards laboratories. The ratio of the candle power of a source to its area is called the luminance of the source; luminances range from about 2,000 candles per square millimetre at the surface of the Sun down to about 3×10^{-6} candle per square centimetre (3×10^{-6} stilb) for the luminous paint on a watch dial. Ordinary outdoor scenes in daylight have an average luminance of several hundred candles per square foot. The quantity of light flux flowing out from a source is measured in lumens, the lumen being defined as the amount of flux radiated by a small "point" source of one candle power into a cone having a solid angle of one steradian. When light falls upon a surface it produces illumination (i.e., illuminance), the usual measure of illuminance being the foot-candle, which is one lumen falling on each square foot of receiving surface.

It is often important to be able to calculate the brightness of an image formed by an optical system, because photographic emulsions and other light receptors cannot respond satisfactorily if the light level is too low. The problem is to relate the luminance of an object with the illuminance in the image, knowing the transmittance and aperture of the optical system. A small area A of a plane object having a luminance of B candles per square unit will have a normal intensity of AB candles. This source radiates light into a cone of semi-angle U, limited, for example, by the rim of a lens. The light flux (F) entering the cone can be found by integration to be,

$$F = \pi AB \sin^2 U \text{ lumens}.$$

If the object luminance is expressed as B_L lamberts, the lambert being an alternative luminance unit equal to $1/\pi$ (i.e., 0.32) candle per unit area, the flux (F) is,

$$F = AB_L \sin^2 U \text{ lumens}$$

because there are π times as many lamberts in a given luminance as there are candles per unit area.

A fraction t of this flux finds its way to the image, t being the lens transmittance, generally about 0.8 or 0.9 but less if a mirror is involved. The area of the image is Am^2, in which m, the magnification, is given by,

$$m = \frac{\sin U}{\sin U'},$$

Hence, the image illuminance (E) is,

$$E = (t\pi \text{AB} \sin^2 U) \div (A \sin^2 U / \sin^2 U') = t\pi B \sin^2 U' \text{ or}$$

$$E = tB_L \sin^2 U'.$$

The image illuminance thus depends *only* on the luminance of the source and the cone angle of the beam proceeding from the lens to the image. This is a basic and most important relation underlying all calculations of image illuminance.

It is often more convenient to convert the angle U' into other better known quantities, such as the *f*-number of the lens and the image magnification. The relation here is,

$$\sin U' = \frac{1}{2(f - number)(1 + m/m_P)}.$$

The *f*-number of the lens is defined as the ratio of the focal length to the diameter of the entrance pupil; *m* is the image magnification; and *mp* is the pupil magnification—i.e., the diameter of the exit pupil divided by the diameter of the entrance pupil. Combining equations $E = (t\pi \text{AB} \sin^2 U) \div (A \sin^2 U / \sin^2 U') = t\pi B \sin^2 U'$ or $E = tB_L \sin^2 U'$,

and $\sin U' = \dfrac{1}{2(f - number)(1 + m / m_P)}$ gives,

$$\text{Image illuminance} = E = \frac{\pi t B}{4(f - \text{number})^2 (1 + m / m_p)^2}.$$

As an example in the use of this relation, if it is supposed that an *f*/2 lens is being used to project an image of a cathode-ray tube at five times magnification, the tube luminance being 5,000 foot-lamberts (1.7 candles per square centimetre), the lens transmittance is 0.8, and the pupil magnification is unity. Then the image illuminance will be:

$$E = \frac{0.8 \times 5{,}000}{4 \times 4 \times 36} = 6.9 \text{ foot-candles}.$$

The image is very much less bright than the object, a fact that becomes clear to anyone attempting to provide a bright projected image in a large auditorium.

Distribution of Illumination over an Image

So far only the illumination at the centre of an image has been considered, but the distribution of illumination over a wide field is often important. In the absence of any lens, the small plane source already considered radiates in a direction inclined at an angle ϕ to the axis with an intensity *AB* cos ϕ. This light has to travel farther than the axial light to reach a screen, and then it strikes the screen at another angle ϕ. The net result is that the oblique illumination on the screen is smaller than the axial illumination by the factor $\cos^4 \phi$.

The same law can be applied to determine the oblique illumination due to a lens, assuming a uniform extended diffusing source of light on the other side of the lens. In this case, however, the exit pupil will not in general be a perfect circle because of possible distortion of the iris by that part of the optical system lying between the iris and the image. Also, any mechanical vignetting in the lens will make the aperture noncircular and reduce still further the oblique illumination. In a camera this reduction in oblique illumination results in darkened corners of the picture, but, if the reduction in brightness is gradual, it is not likely to be detected because the eye adapts quickly to changing brightness as the eyes scan over the picture area. Indeed, a 50 percent drop in brightness between the centre and corners of an ordinary picture is scarcely detectable.

Visual Brightness

The apparent brightness of things seen by the eye follows the same laws as any other imaging system, because the apparent brightness is measured by the illuminance in the image that is projected on the retina. The angle U' in equation ($E = (t\pi \text{AB} \sin^2 U) \div (A \sin^2 U / \sin^2 U') = t\pi B \sin^2 U'$ or $E = tB_L \sin^2 U'$) inside the eye is determined by the size of the pupil of the eye, which varies from about one millimetre to about eight millimetres, depending on the brightness of the environment. Apart from this variation, retinal illuminance is directly proportional to object luminance, and objects having the same luminance appear equally bright, no matter at what distance they are observed.

From this argument, it is clear that no visual instrument, such as a telescope, can possibly make anything appear brighter than when viewed directly. To be sure, a telescope having a large objective lens accepts more light from an object in proportion to the area of the lens aperture, but it magnifies the image area in the same proportion; so the increased light is spread over an increased area of the retina, and the illuminance remains unchanged. Actually, the telescopic view is always dimmer than the direct view because of light losses in the telescope due to glass absorption and surface reflections and because the exit pupil of the telescope may be smaller than the pupil of the eye, thus reducing the angle U'.

The case of a star being observed through a telescope is quite different, because no degree of magnification can possibly make a star appear as anything other than a point of light. Hence, star images appear brighter in proportion to the area of the telescope objective (assuming that the exit pupil is larger than the eye pupil), and the visibility of a star against the sky background is thus improved in proportion to the square of the diameter of the telescope objective lens.

GEOMETRICAL OPTICS

In many situations, the wavelengths of the light being studied are very small compared to the dimensions of the equipment used to study the light, and the photon energies are much smaller than the energy sensitivity of the equipment. Under these conditions we can make an approximation called geometrical optics or ray optics.

Consider the wavelength scale of light waves. Wavelengths in the middle of the visible band are on the order of 500 nm. So a laser beam with a diameter of 1 mm has a diameter of 2000 wavelengths. Individual atoms in a solid are separated by distance on the order of 0.1 nm. So, as far as visible light is concerned, matter is quasi-continuous.

If the wavelengths of the light become comparable to the dimensions of the equipment, but the photon energies are still negligibly small, then we study optical phenomena using the classical theory of radiation, or wave optics. Wave optics contains all of ray optics, but the mathematical treatment is much more involved. When the photon energies become comparable to the sensitivity of the equipment, we must study the phenomena using quantum optics.

We define a ray as the path along which light energy is transmitted from one point to another in an optical system. The basic laws of geometrical optics are the law of reflection and the law of refraction. If not being reflected or refracted, a light ray travels in a straight line. The optical path length of a ray traveling from point A to point B is defined as c times the time it takes the ray to travel from A to B. Assume a ray travels a distance d_1 in a medium with index of refraction n_1 and a distance d_2 in a medium with index of refraction n_2. The speed of light in a medium with index of refraction n is c/n. The travel time from A to B therefore is:

$$t = n_1 d_1/c + n_2 d_2/c.$$

The optical path length is OPL = $n_1 d_1 + n_2 d_2$.

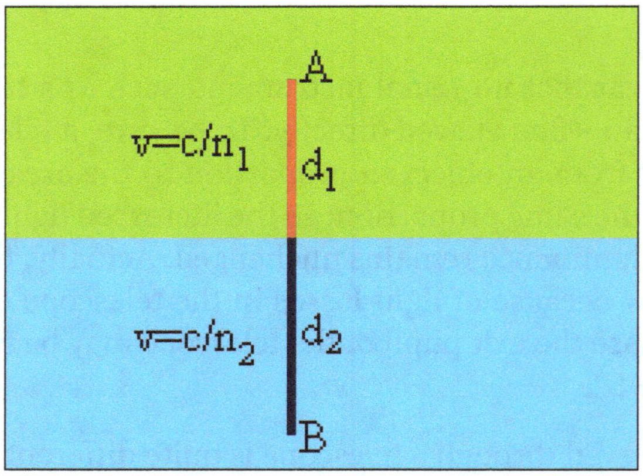

Table: The index of refraction for various media.

Absolute vacuum	n = 1	v = c
Air	n = 1.0003	v = 0.9997 c
Water	n = 1.33	v = 0.75 c
Glass	1.4 < n < 1.8	0.56 c < v < 0.71 c
Diamond	n = 2.4	v = 0.42 c
Silicon	n = 3.5	v = 0.29 c

The subject of geometrical optics starts with the laws of refraction and reflection for transparent media. These laws are used to discover the properties of various optical systems, which may contain any number of curved refracting and reflecting surfaces. In principle one could solve any optical problem by the exact application of the basic laws. However, it is possible to derive some very useful general results by using approximate forms of the laws, which treat only rays, called paraxial rays, which make small angles with respect to an optical axis. Often the treatment is confined to spherical surfaces. This approximate form of the basic laws as applied to paraxial rays in an optical system consisting of spherical surfaces is called the first order theory and constitutes a major part of the subject of geometrical optics.

PHYSICAL OPTICS

In physics, physical optics, or wave optics, is the branch of optics that studies interference, diffraction, polarization, and other phenomena for which the ray approximation of geometric optics is not valid. This usage tends not to include effects such as quantum noise in optical communication, which is studied in the sub-branch of coherence theory.

Waves and Diffraction

Mathematical Description of a Wave

A wave is a periodically repeating disturbance in the value of some quantity. In the case of light this oscillation is of the electric or magnetic field strength or amplitude.

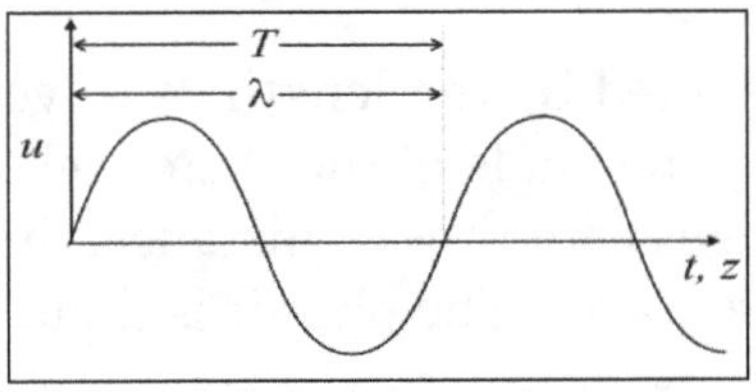

$$u = u_0 \cos(kz - \omega t + \alpha) \qquad \text{or} \qquad u = u_0 e^{i\alpha} e^{i(kz - \omega t)}$$

Where u is the amplitude, ω is the angular frequency and *k* is the wave number for the wave oscillation of period T and wavelength λ. The wave number *k* is a scalar but is sometimes used as a vector, k, to indicate the direction of the wave propagation.

ωt: Phase change with time, $\omega = 2p / T$.

kz: Phase change with distance, $k = 2p / \lambda$.

α: Arbitrary initial phase.

This convention for a wave travelling from left to right i.e. in the positive z direction follows that used in Quantum Mechanics to describe wave functions.

Interference

The addition of amplitudes from two sources gives interference e.g. Young's slits:

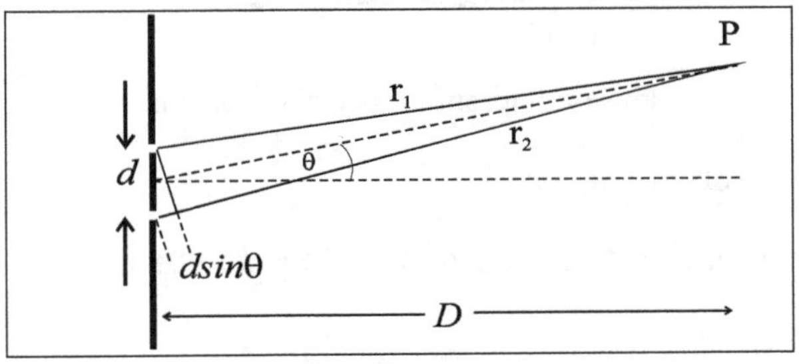

Young's slits.

We consider two slits separated by d illuminated by monochromatic plane waves Amplitude up at a point u_p large distance, *D*, from the slits.

$$u_p = \frac{u_0}{r_1} e^{i(kr_1 - \omega t)} + \frac{u_0}{r_2} e^{i(kr_2 - \omega t)}$$

The factors of $1/r_1$ and $1/r_2$ account for the fall-off in amplitude according to the inverse square law for the intensity of the wave.

Putting $(r_1 - r_2) = d\sin\theta, r_1 \sim r_2 = r,$ the intensity is:

$$I_p = 4\left(\frac{u_0}{r}\right)^2 \cos^2(\frac{1}{2}kd\sin\theta).$$

Phasors

The amplitude of a wave is represented by the length of a "vector" on an Argand diagram. The phase of the wave is represented by the angle of the vector relative to the Real axis of the Argand diagram. So that, as the phase changes with time or distance, the direction of the phasor will rotate in the plane of the real and imaginary axes. The phasor is then: $ue^{i\delta}$.

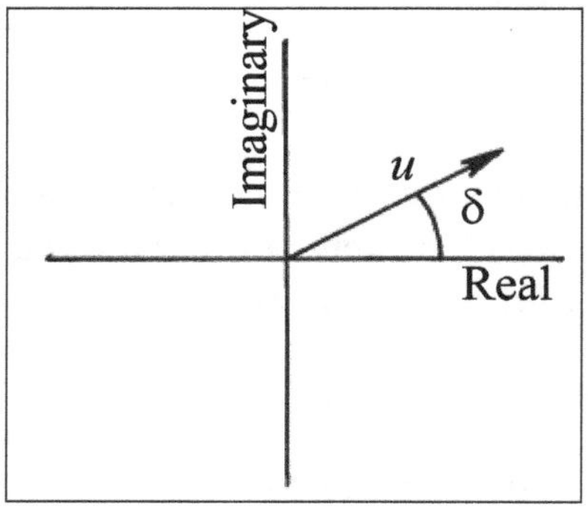

Phasor diagram.

Example: Young's slits.

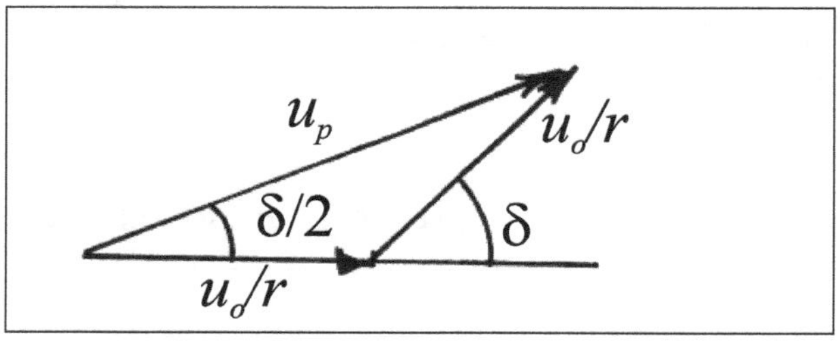

Phasor diagram for two slit problem.

Amplitude from each slit on screen: u_0 / r.

Phase difference δ, owing to path difference $d\sin\theta : \delta = kd\sin\theta$.

Resultant amplitude is then,

$$u_p = 2\frac{u_0}{r}\cos(\delta / 2)$$

The intensity is therefore:

$$I_p = 4\left(\frac{u_0}{r}\right)^2 \cos^2(\frac{1}{2}kd\sin\theta).$$

Diffraction from a Finite Slit

A monochromatic plane wave incident on aperture of width a, is observed in a plane at large distance D from aperture. The amplitude in the plane of aperture is u_o per unit length.

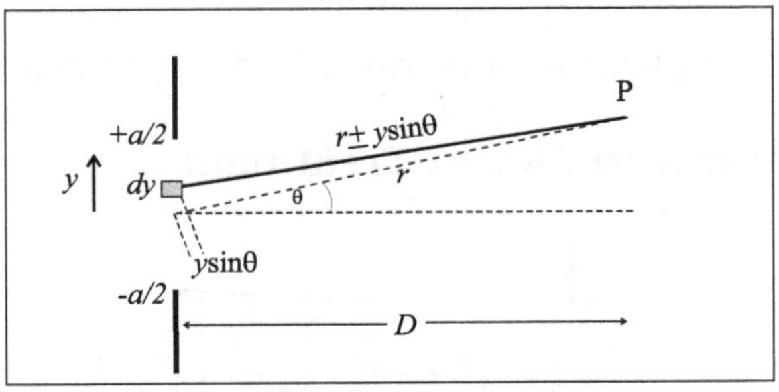

Contributions to amplitude at P from elements dy in slit.

An infinitesimal element of length dy at position y contributes at P an amplitude:

$$\frac{u_0 dy}{r} e^{i\delta(y)}$$

The phase factor $\delta(y) = k(r \pm ysin\theta)$.

The total amplitude at P arising from all contributions across the aperture:

$$u_p = \frac{u_0}{r} e^{ikr} \int_{-a/2}^{a/2} e^{ik\sin\theta . y} dy$$

The intensity is then:

$$I_p = I(0)\sin c^2 \beta$$

where,

$$\beta = \frac{1}{2} ka \sin\theta$$

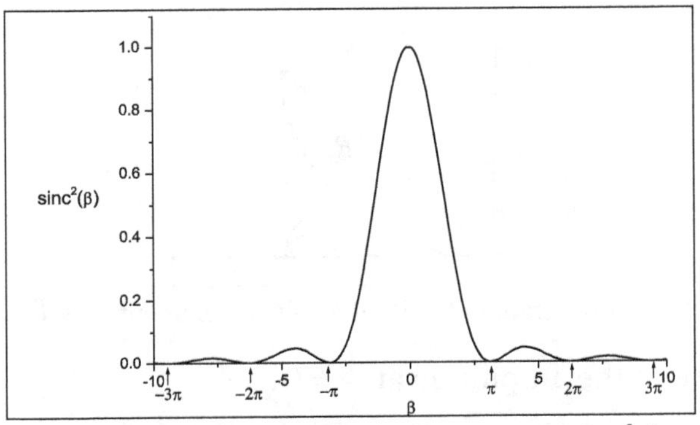

Intensity pattern from single slit, $I_p = I(0)\sin c^2 \beta$.

The first minimum is at $\beta = \pi$, $\qquad 2\pi = \frac{2\pi}{\lambda} a \sin\theta$

Hence angular width θ of diffraction peak is:

$$\theta = \frac{\lambda}{a}$$

This result represents the basic physics of all wave diffraction phenomena.

Diffraction from a Finite Slit: Phasor Treatment

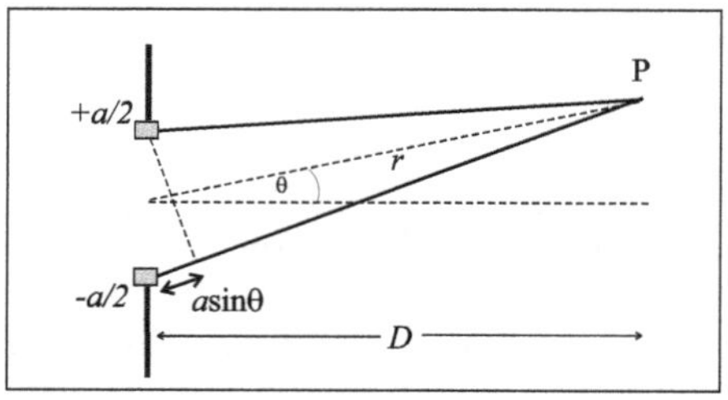

Construction showing elements at extreme edges of aperture contributing first and last phasors.

On axis, $\theta = 0$ the phasor elements sum to R_p Off axis, $\theta \neq 0$ successive phase shifts between adjacent phasors bend the phasor sum to form a section of a regular polygon.

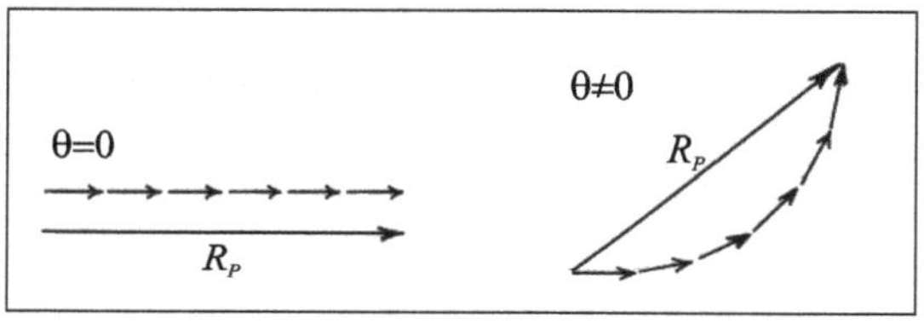

Phasor diagram for finite slit showing resultant R_p for $\theta = 0$ and $\theta \neq 0$.

The phase difference between first and last phasors for $\theta \neq 0$ is:

$$\delta = ka\sin\theta$$

In the limit as the phasor elements $\to 0$ the phasors form an arc of a circle of radius R. The length of the arc is R_o and the length of the chord representing the resultant is R_p.

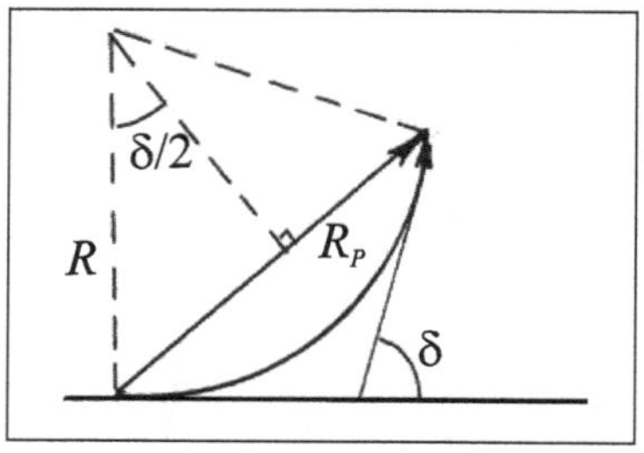

Phasor diagram in the limit as phasor elements $\to 0$.

The amplitude at θ relative to the amplitude at $\theta = 0$:

$$\frac{\textit{length of chord}}{\textit{length of are}} = \frac{2R\sin(\delta/2)}{R.\delta} = \sin c(\delta/2)$$

Then the intensity at θ:

$$\mathrm{I}(\theta)=\mathrm{I}(0)\ \sin c^2\ (\delta/2)=\mathrm{I}(0)\ \sin c^2\ \beta$$

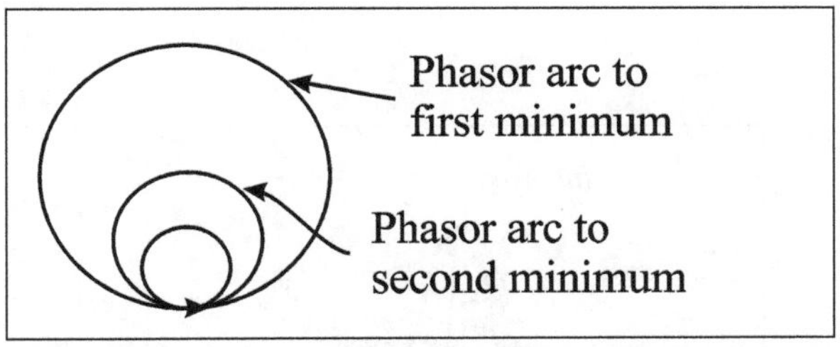

Phasor diagram showing mimina for increasing phase shift δ between extremes of slit as θ increases.

The first minimum occurs when the phasor arc bends to become a full circle i.e. the phase difference between first and last phasor elements $\delta = 2\pi$, the angular width is:

$$\theta = \frac{\lambda}{a}.$$

Diffraction in 2 Dimensions

Recall that the amplitude resulting from a plane wave illumination of an aperture of the form of a slit of width a in the y - direction:

$$u_p = \frac{u_0}{r} e^{ikr} \int_{-a/2}^{a/2} e^{ik\sin\theta.y} dy$$

Consider the aperture to have a width b in the x-direction, then the angular variation of the diffracted amplitude in the x-direction is :

$$u_p = \frac{u_0}{r} e^{ikr} \int_{-b/2}^{b/2} e^{ik\sin\phi.x} dx$$

In 2 - D we have : $u_p \propto e^{ikr} \int_{-b/2}^{b/2} \int_{-a/2}^{a/2} u(x, y) e^{ik(\sin\phi.x + \sin\theta.y)} dxdy$

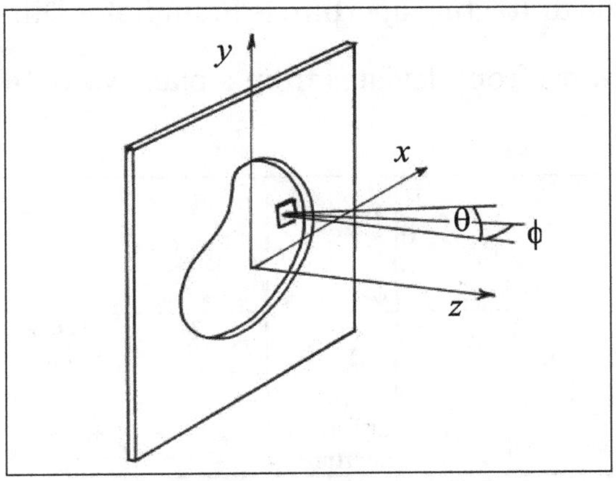

General 2-D aperture in x,y plane.

where, $u(x, y)$ is the amplitude distribution function for the aperture. For a circular aperture of diameter a the diffraction pattern is a circular Bessel Function. The angular width to the first minimum is:

$$\theta = 1.22\frac{\lambda}{a}$$

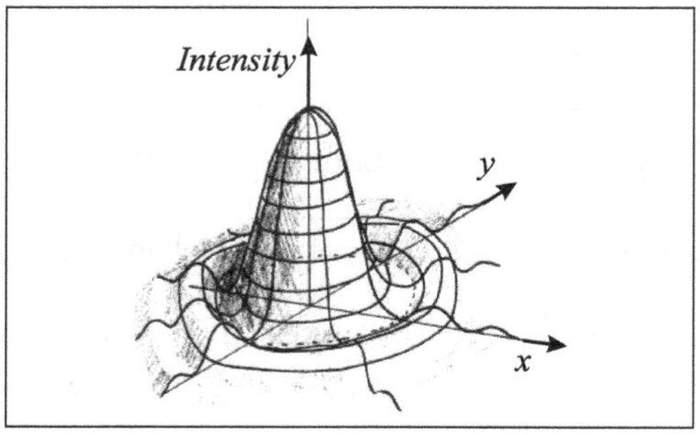

Point Spread Function for circular aperture.

A point source imaged by a lens of focal length f and diameter a gives a pattern with a minimum of radius $r = f\lambda / a$. This is the Point Spread Function or instrument function analogous to the impulse function of an electrical circuit giving its response to a δ-function impulse.

Fraunhofer Diffraction

So far we have considered diffraction by:

- Apertures or slits illuminated by plane waves.
- Observation at a large distance where the phase difference between contributions from secondary sources in the diffracting plane separated by y is given to a good approximation by:

  $\delta = k \ \sin\theta . y$

These are special cases where the phase difference δ is a linear function of the position y in the diffracting aperture.

"A diffraction pattern for which the phase of the light at the observation point is a linear function of position for all points in the diffracting aperture is Fraunhofer Diffraction".

By linear we mean that the wave front deviates from a plane wave by less than $\lambda / 20$ across the diffracting aperture.

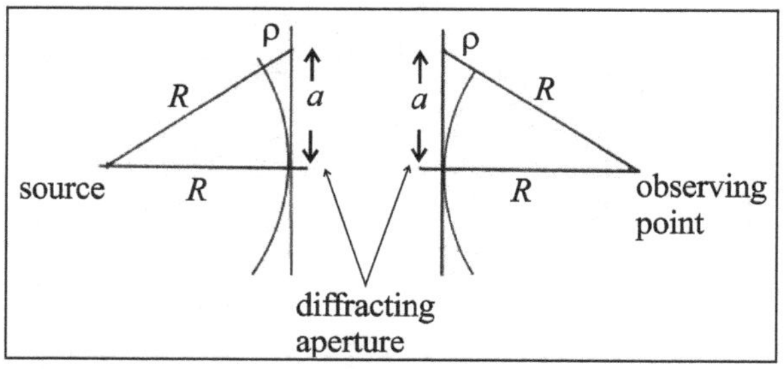

Wavefronts incident on and exiting from a plane aperture.

$$(R+\rho)^2 = R^2 + a^2$$
$$for\ \rho \leq \lambda/20 \text{ and } R \approx \sim 10a^2/\lambda$$

Alternatively,

"Fraunhofer diffraction is the diffraction observed in the image plane of an optical system."

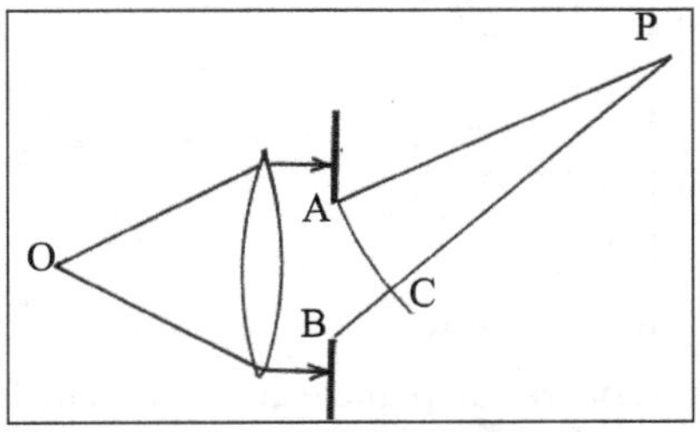

Fraunhofer condition for plane waves: image is at infinity as source is at focal length from lens.

Consider a point source at the focal point of a lens so that collimated light (plane waves) are incident on an aperture behind the lens. The image of the source is at ∞. Fraunhofer Diffraction will be observed at P if $BC \leq \lambda/20$.

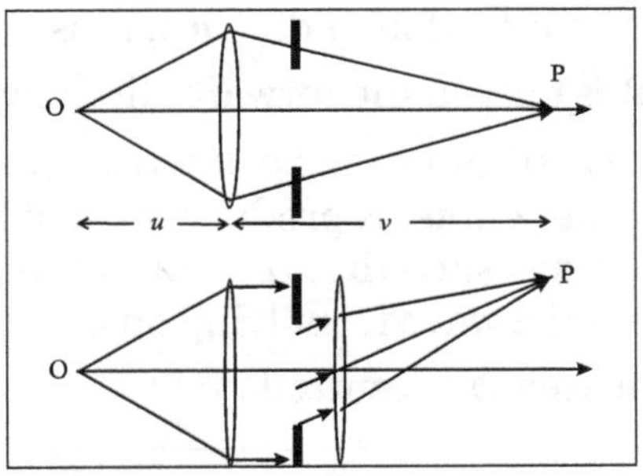

Fraunhofer diffraction observed in the image plane of a lens.

If the observation point P lies in the image plane of the lens so that curved wavefronts converge from the lens to P then no plane waves are involved. The lens and diffracting aperture however can be replaced by an equivalent system where diffraction of plane waves occurs. Note however that this means plane waves are not necessary to observe Fraunhofer diffraction. The key criterion is that the phase varies linearly with position in the diffracting aperture.

A further consequence of noting that Fraunhofer diffraction is observed in the image plane is that the position of the aperture is not important.

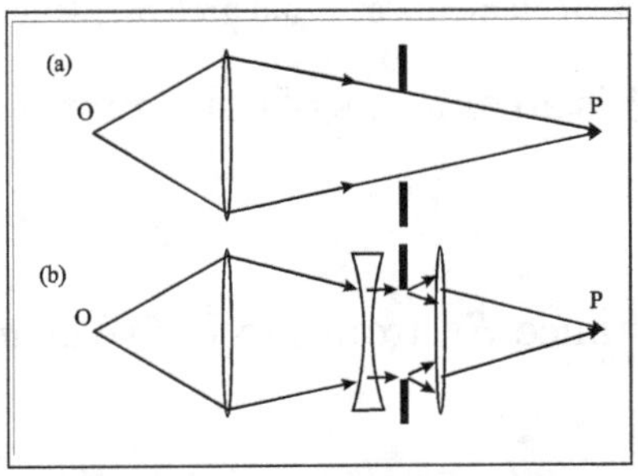

Equivalent lens system showing that Fraunhofer diffraction is independent of position of aperture.

Diffraction and Wave Propagation

Consider a plane wave surface at -z. This reproduces itself at a second plane z = 0. Huygens secondary sources in the wave front radiate to a point P in the second plane.

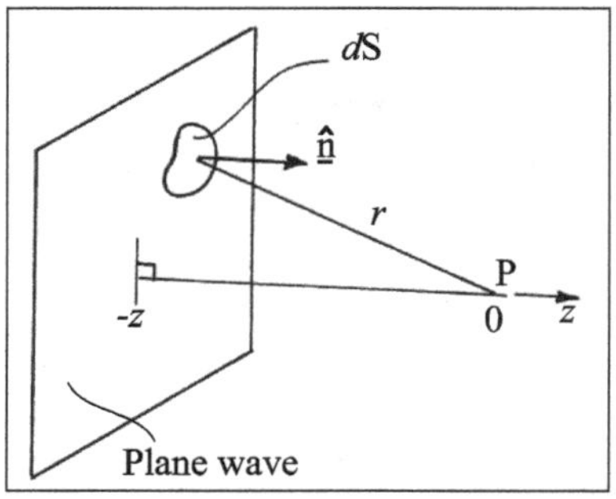

Huygens secondary sources on plane wave at -z contribute to wave at P.

The amplitude at P is the resultant of all contributions from the plane at −z.

$$u_p = \alpha \int \frac{u_0 dS}{r} \eta(\text{n, r}) e^{ikr}$$

where, u_o is the amplitude from element of area dS. $\eta(\text{n, r})$ is the obliquity factor - this accounts for the fact that the wave propagates only in the forward direction. n is a unit vector normal to the wave front and α is a proportionality constant - to be determined. We determine α by a self-consistency argument i.e. the plane wave at -z must reproduce itself at z = 0. We consider the amplitude at α point P a distance q from the wave such that $q = m\lambda$ where m is an integer and $m >> 1$ i.e. P is a large distance from O, a point on the wave front lying on a normal through P. We now construct elements of the wavefront of equal area δA centred on O.

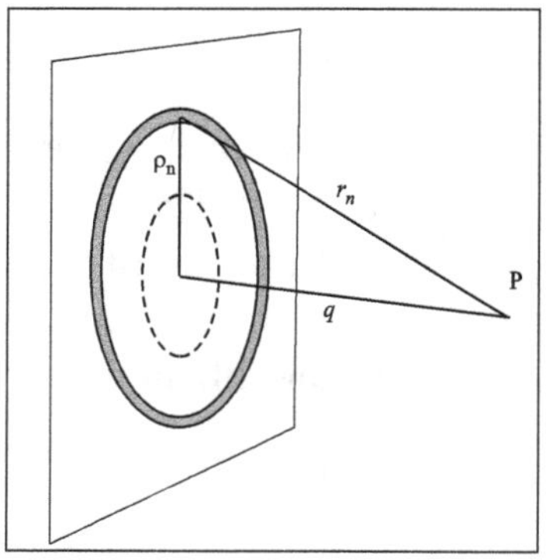

Construction of elements of equal area on plane wavefront.

The first element is a circle, the nth is an annulus of outer radius ρ_n:

$$\pi(\rho_{n+1}^2 - \rho_n^2) = \delta A$$

Consequently the difference in distance δr from successive elements to P is constant:

$$\delta r = r_{n+1} - r_n \cong \frac{\delta A}{2\pi q}$$

Therefore the phase difference between waves from successive elements is also constant:

$$\delta\phi = \frac{\delta A}{\lambda q}$$

Hence we may treat contributions from each element of the wavefront as a Phasor. We ignore, for the moment, the small difference in amplitude at P between successive elements arising from the small increase in distance r_n as ρ_n increases. We also ignore the small change in $\eta(n, r)$.

Add contributions of elements (phasors) until the last phasor added is π out of phase with the first. The area of the wavefront covered by these elements is the First Half Period Zone, 1st HPZ.

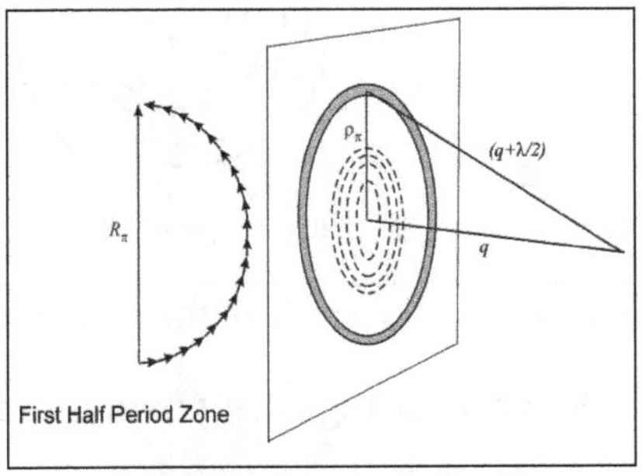

Construction of the First Half Period Zone.

The difference in path-length from the outer element of the 1st HPZ to P and from O to P is $\lambda/2$.

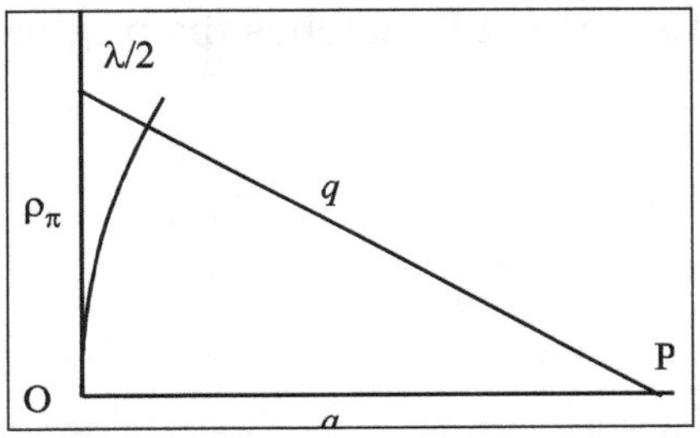

Phase shift of $\lambda/2$ arises at edge of 1st HPZ.

The radius of the 1st HPZ is $\rho\pi$ is given by:

$$\rho_\pi^2 = \lambda q$$

Recalling our diffraction integral we write the contribution to the amplitude at P from the 1st HPZ:

$$\alpha \frac{u_0 \pi \rho_\pi^2}{q} = \alpha u_o \pi \lambda$$

From the phasor diagram, the amplitude from the 1st HPZ is the length of the phasor.

The resultant R_π is then the diameter of the circle of which the phasor arc defines half the circumference:

$$\frac{\pi R_\pi}{2} = \alpha u_o \pi \lambda$$

The resultant phasor lies along the imaginary axis so:

$$R_{\pi} = 2i\alpha u_{o}\lambda$$

Add further elements until the final phasor is in phase with the first i.e. a phase difference of 2π. The area of the wavefront now defines the first Full Period Zone 1st FPZ. The resultant from the 1st FPZ is not exactly zero owing to the term 1/r (inverse square law for intensity) and the obliquity factor $\eta(\text{n, r})$. Adding further elements gives a slow spiral.

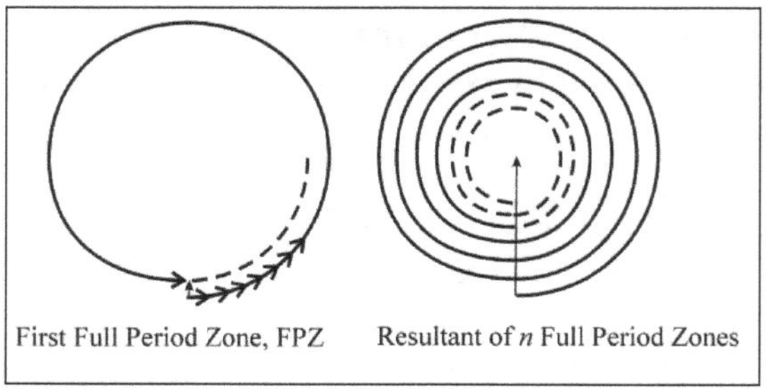

As $n \to \infty$ resultant of zones tends to half the resultant of the 1st HPZ.

Adding contributions from the whole wave (integrating over infinite surface) gives resultant equal to ½ the 1st HPZ. Therefore:

$$R_{\infty} = i\alpha u_{0}\lambda$$

Self-consistency demands that this wave at P matches the original wave at O:

$$u_{0} = i\alpha u_{0}\lambda$$

$$\therefore \alpha = -\frac{i}{\lambda}$$

Hence,

$$u_{p} = -\frac{i}{\lambda}\int \frac{u_{o}dS}{r}\eta(\text{n, r})e^{ikr}$$

This is the Fresnel-Kirchoff diffraction integral.

QUANTUM OPTICS

Quantum optics is primarily an area of physics which uses a combination of semi-classical physics and quantum mechanics principles to investigate and manipulate how photons of light interact with matter, and the phenomena which can be produced, at the subatomic level. This is how you would explain quantum optics in its broadest sense. However, whilst some of the most prominent applications is lasers and quantum computing, there has been a lot of research into the fundamental principles of how photons behave at this level, and this has helped to realize many different subsets and phenomenon within quantum optics which contribute heavily to the realization of the physical applications.

Coincidence Correlation

Coincidence correlation is an area of quantum optics that is used to see if someone is observing a single quantum system. This is done by assuming that a single system can only emit one photon at a time and observes (via a photodetector) the quantum system as a single photon emitter. If it is found that more than detector observes the source, then the likelihood is that it is not a one photon system and is unlikely to be a single quantum system. It is a fundamental process that enables someone to determine the presence of a single quantum system, i.e. it is a test rather than an application, but it can be used in conjunction with other quantum optic application areas.

One example is with quantum entanglement. Coincidence correlation can be used to prove or disprove the correlations with a quantumly entangled network and will employ a combination of optical polarizers and photodetectors to filter quantum states and determine if there is correspondence at both ends of the entangled pair.

Quantum Entanglement

Quantum entanglement is a phenomenon that occurs between quantum systems, where the components of each quantum system become one and indescribable from each other, i.e. instead of two separate quantum states, the whole system becomes one quantum network state. The types of components which can experience this phenomenon include electrons, photon, atoms and molecules. This extends to long-range distances, and the measurement of one part of the quantum system enables the properties of the corresponding particle in the quantum system to be revealed.

The different properties that can be revealed at different ends of an entangled network include position, momentum, spin and polarization. In many cases, one of the quantum particles is described as a superimposition with an indefinite value for the entangled particle. However, if one of these particles is measured, it can provide definite value for the corresponding pair. Quantum entanglement is often utilized in quantum computing applications.

Quantum Teleportation

Quantum teleportation is another phenomenon that has a lot of use in quantum computing, as well as quantum communications, and is closely related to quantum entanglement. Quantum teleportation is the process by which the information held within a qubit can be transported from one location to another, without the qubit itself being transported.

For those who don't know, a qubit, otherwise known as a quantum bit, is the building block of many quantum networks, especially in quantum information processing applications and can adopt a 0 value, a 1 value or a superimposed 0 or 1 value. This means that qubits can perform quantum operations in more than one value simultaneously.

Quantum Information Processing

Quantum information processing, i.e quantum computing, is a computing process (and memory storage) that relies on qubits as opposed to binary bits. The ability to of qubits to superimpose, compared to binary bits that adopt a 0 or 1 value only, enables simultaneous operations to occur (several quantum systems can be operated in parallel), and in turn allows quantum computers to be much faster than their classical counterparts. The qubits within a quantum computer stores information quantum mechanically by utilizing the ½ spin state of electrons (up and down) and the polarization of photons (horizontal and vertical) within the quantum network. This correlates to a positional arrangement that can be identified when the qubits are entangled, and so long as the computer can control the spin operation and the interactions between electron spins, the readout can measure the single spin states and bulk spin resonance of each quantum network to determine the information contained within.

Quantum Communication

Quantum communication is an area which is closely related to quantum information processing but is more to do with quantum cryptography than it is computing—such as quantum key distribution. Quantum key distribution uses quantum mechanics to perform cryptographic tasks or break encrypted systems. Quantum key distribution works when two people use a communication system that utilizes single photons, which are randomly polarized, to transmit a series of random number sequences. The randomness of the polarization is generated by using quantum optics. These sequences act as the keys in the cryptographic system and the system uses both a classical channel and a quantum channel to connect the communication points.

The qubits are sent over the quantum channel and the classical channel performs classical operations and can be used to see if anyone is trying to hack the system. Because the information is transmitted via the quantum network, and not the classical channel, the classical channel can be hacked but no information will be obtained. However, because the signals under normal conditions are correlated, any correlation imperfections (due to a hack) between the classical network and the quantum network will be detected by the receiver and can be used to determine when a hack has been attempted.

RESOLUTION

Resolution is a measurement of an imaging system's ability to reproduce object detail, and can be influenced by factors such as the type of lighting used, the pixel size of the sensor, or the capabilities of the optics. The smaller the object detail, the higher the required resolution.

Dividing the number of horizontal or vertical pixels on a sensor into the size of the object one wishes to observe will indicate how much space each pixel covers on the object and can be used to estimate resolution. However, this does not truly determine if the information on the pixel is distinguishable from the information on any other pixel.

As a starting point, it is important to understand what can actually limit system resolution. An example can be shown in figure, a pair of squares on a white background. If the squares are imaged onto neighboring pixels on the camera sensor, then they will appear to be one larger rectangle in the image rather than two separate squares. In order to distinguish the squares, a certain amount of space is needed between them, at least one pixel. This minimum distance is the limiting resolution of the system. The absolute limitation is defined by the size of the pixels on the sensor as well as the number of pixels on the sensor.

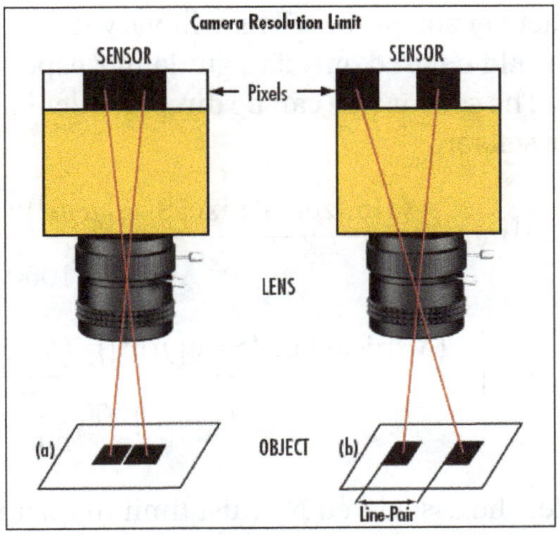

Resolving Two Squares. If the space between the squares is too small (a) the camera sensor will be unable to resolve them as separate objects.

The Line Pair and Sensor Limitations

The relationship between alternating black and white squares is often described as a line pair. Typically, the resolution is defined by the frequency measured in line pairs per millimeter (lp/mm). A lens's resolution is unfortunately not an absolute number. At a given resolution, the ability to see the two squares as separate entities will be dependent on grey scale level. The bigger the separation in the grey scale between the squares and space between them, the more robust is the ability to resolve the squares. This grey scale separation is known as contrast (at a specified frequency). The spatial frequency is given in lp/mm. For this reason, calculating resolution in terms of lp/mm is extremely useful when comparing lenses and for determining the best choice for given sensors and applications.

The sensor is where the system resolution calculation begins. By starting with the sensor, it is easier to determine what lens performance is required to match the sensor or other application requirements. The highest frequency which can be resolved by a sensor, the Nyquist frequency, is effectively two pixels or one line pair. The resolution of the sensor, also referred to as the image space resolution for the system, can be calculated by multiplying the pixel size in μm by 2 (to create a pair), and dividing that into 1000 to convert to mm:

$$\text{Sensor Resolution}\left[\frac{\text{lp}}{\text{mm}}\right] = \text{Image Space Resolution}\left[\frac{\text{lp}}{\text{mm}}\right] = \frac{1000\frac{\mu\text{m}}{\text{mm}}}{2\times\text{Pixel Size}[\mu\text{m}]}$$

Sensors with larger pixels will have lower limiting resolutions. Sensors with smaller pixels will have higher limiting resolutions.

With this information, the limiting resolution on the object to be viewed can be calculated. In order to do so, the relationships between the sensor size, the field of view, and the number of pixels on the sensor need to be understood.

Sensor size refers to the size of a camera sensor's active area, typically specified by the sensor format size. However, the exact sensor proportions will vary depending on the aspect ratio, and the nominal sensor formats should be used only as a guideline, especially for telecentric lenses and high magnification objectives. The sensor size can be directly calculated from the pixel size and the number of active pixels on the sensor.

$$\text{Horizontal Sensor Dimension}[\text{mm}] = \frac{(\text{Horizontal Pixel Size}[\mu\text{m}]) \times (\#\text{ of active horizontal pixels})}{1000\frac{\mu\text{m}}{\text{mm}}}$$

$$\text{Vertical Sensor Dimension}[\text{mm}] = \frac{(\text{Vertical Pixel Size}[\mu\text{m}]) \times (\#\text{ of active vertical pixels})}{1000\frac{\mu\text{m}}{\text{mm}}}$$

Table: As pixel sizes get smaller the associated Nyquist limit in lp/mm rises proportionally.

Pixel Size (μm)	Associated Nyquist Limit (lp/mm)
1.67	299.4
2.2	227.3
3.45	144.9
4.54	110.1
5.5	90.9

OPTICAL ILLUSION

Illusions are special perceptual experiences in which information arising from "real" external stimuli leads to an incorrect perception, or false impression, of the object or event from which the stimulation comes.

Some of these false impressions may arise from factors beyond an individual's control (such as the characteristic behaviour of light waves that makes a pencil in a glass of water seem bent), from inadequate information (as under conditions of poor illumination), or from the functional and structural characteristics of the sensory apparatus (e.g., distortions in the shape of the lens in the eye). Such visual illusions are experienced by every sighted person.

Another group of illusions results from misinterpretations one makes of seemingly adequate sensory cues. In such illusions, sensory impressions seem to contradict the "facts of reality" or fail to report their "true" character. In these instances the perceiver seems to be making an error in processing sensory information. The error appears to arise within the central nervous system (brain and spinal cord); this may result from competing sensory information, psychologically meaningful distorting influences, or previous expectations (mental set). Drivers who see their own headlights reflected in the window of a store, for example, may experience the illusion that another vehicle is coming toward them even though they know there is no road there.

Types of Illusory Experiences

Stimulus-distortion Illusions

This type of illusory sense perception arises when the environment changes or warps the stimulus energy on the way to the person, who perceives it in its distorted pattern.

Auditory Phenomena

A common phenomenon is the auditory impression that a blowing automobile horn changes its pitch as it passes an observer on a highway. This is known as the Doppler effect, for Christian Doppler, an Austrian physicist, who in 1842 noted that the pitch of a bell or whistle on a passing railroad train is heard to drop when the train and the perceiver are moving away from each other and to grow higher when they are approaching each other. The sound heard is also affected by factors such as a wind blowing toward or away from the person.

Another auditory illusion was described in 1928 by Paul Thomas Young, an American psychologist, who tested the process of sound localization (the direction from which sound seems to come). He constructed a pseudophone, an instrument made of two ear trumpets, one leading from the right side of the head to the left ear and the other vice versa. This created the illusory impression of reversed localization of sound. While walking along the street wearing the pseudophone, he would hear footsteps to his right when they actually came from the left.

When two sources of sound in the same vicinity emit sound waves of slightly different frequencies (i.e., vibrations per second), there will come intervals when waves from both sources arrive at the ear in phase (simultaneously) and produce the experience of a combined, louder sound. These intervals of combined sound will be perceived as "beats," or periodic alternations of sound intensity. When such auditory beats occur too rapidly to be discriminated, a harsh, continuous noise, commonly called interference, may result. Another case of interference results when two tones sounded together produce a subjectively heard third tone. When this third tone is lower in pitch than the original two, it is called a difference tone; i.e., its frequency is the difference between the frequencies of the two original tones. When the third tone is higher, it is called a summation tone; i.e., its frequency is the sum of the frequencies of the two original tones. Piano tuners depend in part upon their ability to hear these tones in a reliable way when tightening and loosening the strings in order to reach the correct pitch on the instrument.

Optical Phenomena

Numerous optical illusions are produced by the refraction (bending) of light as it passes through one substance to another in which the speed of light is significantly different. A ray of light passing from one transparent medium (air) to another (water) is bent as it emerges. Thus, the pencil standing in water seems broken at the surface where the air and water meet; in the same way, a partially submerged log in the water of a swamp gives the illusion of being bent.

Rainbows also result from refraction. As the sun's rays pass through rain, the droplets separate (refract) the white light into its component colours. As rays of white light from any source pass through a prism, they are refracted to give the appearance of a spectrum of colour, as in the rainbow of a summer morning. Another illusion that depends on atmospheric conditions is a mirage, in which, for example, the vision of a pool of water is created by light passing through layers of air above the heated surface of a highway. In effect, cooler layers of air refract the sun's rays at different angles than do less-dense strata of heated air, giving the appearance of water where there is none; nearby objects may even appear to be reflected in it. Under certain conditions, elaborate mirages that look like cities, forests, or "unidentified flying objects" may appear on the horizon, or ships in a nearby body of water may appear to be plying the sky of a desert.

Optical Illusion: Refraction of light. The refraction (bending) of light as it passes from air into water causes an optical illusion: straws in the glass of water appear broken or bent at the water's surface.

A mirage in a desert in Africa resembles a shimmering pool of water. The illusion is caused by the refraction (bending) of light in layers of air above the desert's hot surface.

Perceiver-distortion Illusions

Some illusions are related to characteristics of the perceiver, namely the functioning of the brain and the senses, rather than to physical phenomena that distort a stimulus. Many common visual illusions are perceptual: they result from the brain's processing of ambiguousor unusual visual information. Other illusions result from the aftereffects of sensory stimulation or from conflicting sensory information. Still others are associated with psychiatric causes.

Visual Perceptual Illusions

When an observer is confronted with a visual assortment of dots, the brain may group the dots that "belong together." These groupings are made on the basis of such things as observed similarity (e.g., red versus black dots), proximity, common direction of movement, perceptual set (the way one is expecting to see things grouped), and extrapolation (one's estimate of what will happen based on an extension of what is now happening).

Closure (a term used in Gestalt psychology) is the illusion of seeing an incomplete stimulus as though it were whole. Thus, one unconsciously tends to complete (close) a triangle or a square that has a gap in one of its sides. While a person watches a movie, closure occurs to fill the intervals between what are really rapidly projected still pictures—giving the illusion of uninterrupted motion.

The "figure-ground" illusion is commonly experienced when one gazes at the illustration of a white vase, the outline of which is created by two black profiles. At any moment, one will be able to see either the white vase (in the centre area) as "figure" or the black profiles on each side (in which case the white is seen as "ground"). The fluctuations of figure and ground may occur even without conscious effort. Seeing one aspect usually excludes seeing the other.

Ambiguous figure seen as either a white vase or two black profiles.

Another example of ambiguity and object reversibility is the Necker cube, which may seem to flip-flop. Some studies have suggested that younger people tend to perceive these reversals more readily than do their elders.

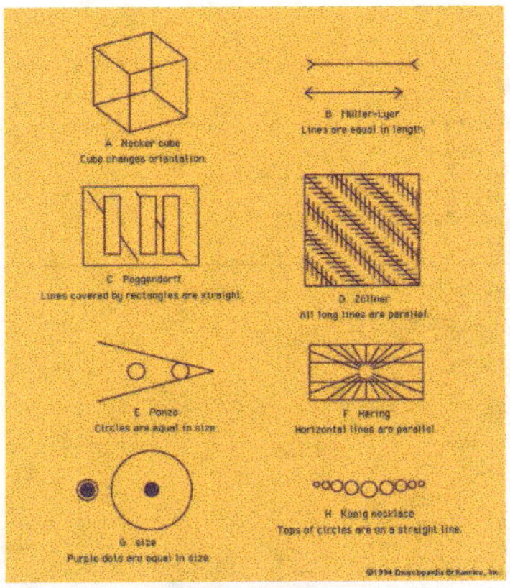

Examples of optical illusions. (A) Cube changes orientation. (B) Lines are equal in length. (C) Lines covered by rectangles are straight. (D) All long lines are parallel. (E) Circles are equal in size. (F) Horizontal lines are parallel. (G) Black dots are equal in size. (H) Tops of circles are on a straight line.

The Müller-Lyer illusion is based on the Gestalt principles of convergence and divergence: the lines at the sides seem to lead the eye either inward or outward to create a false impression of length. The Poggendorff illusion depends on the steepness of the intersecting lines. As obliqueness is decreased, the illusion becomes less compelling. In the Zöllner illusion, the cross-hatching disturbs the perception of parallel lines. A figure seen touching converging lines, as in the Ponzo illusion, seems larger than another figure of the same size placed between the lines where they are farther apart. In a related experience, linear perspectivecreates the illusion that parallel lines or contours (such as railroad tracks) converge as they recede from the viewer.

In studies of visual verticality, experimenters investigated the conditions that determine perception of the "upright." A tilted chair that could be mechanically controlled by the subject was placed in a slanted room containing visual indicators of verticals and horizontals. When various persons were asked to sit in the chair and align themselves in a vertical position, some of the subjects aligned themselves with the "true vertical" determined by gravity, while others experienced the illusion of verticality by aligning themselves with the visual directions they saw in the slanted room. Closing the eyes made "true" alignment easier.

Staring at a single bright spot in an otherwise darkened room creates the illusion that the stationary light is moving (autokinetic effect). One theory to account for this is that the impression is caused by minute eye movements of the observer. The so-called phi phenomenon is an illusion of movement that arises when stationary objects—light bulbs, for example—are placed side by side and illuminated rapidly one after another. The effect is frequently used on theatre marquees to give the impression of moving lights.

Perhaps the best real-life example of a perceptual illusion is the Moon illusion. When the Moon is at the horizon, it appears to be much larger than it does when it is high in the sky. Yet when the Moon is photographed at various points across the sky, all the images on the negatives are the same size. Considerable debate surrounds the source of the Moon illusion. Some explanations have attributed it to the paradoxical idea that the Moon at the horizon seems larger because the brain perceives it as being farther away than the Moon at the sky's zenith. Another explanation is that the lack of distance cues in the night sky causes the eyes to adjust to a near-focus position, which makes the high Moon appear smaller.

BIREFRINGENCE

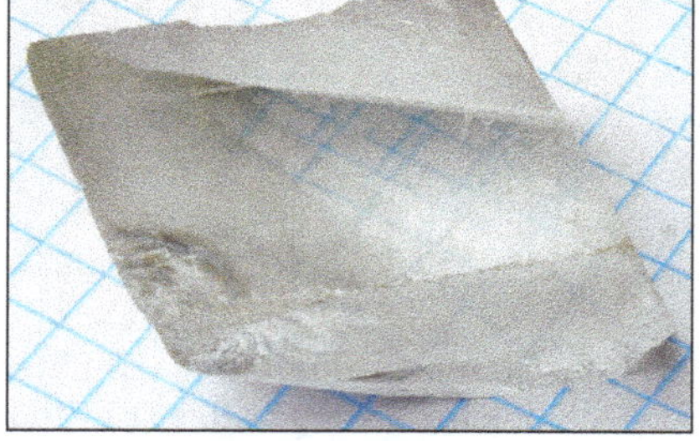

A calcite crystal laid upon a graph paper with blue lines showing the double refraction.

Doubly refracted image as seen through a calcite crystal, seen through a rotating polarizing filter illustrating the opposite polarization states of the two images.

Birefringence is the optical property of a material having a refractive index that depends on the polarization and propagation direction of light. These optically anisotropic materials are said to be birefringent (or birefractive). The birefringence is often quantified as the maximum difference between refractive indices exhibited by the material. Crystals with non-cubic crystal structures are often birefringent, as are plastics under mechanical stress.

Birefringence is responsible for the phenomenon of double refraction whereby a ray of light, when incident upon a birefringent material, is split by polarization into two rays taking slightly different paths. This effect was first described by the Danish scientist Rasmus Bartholin in 1669, who observed itin calcite, a crystal having one of the strongest birefringences. However it was not until the 19th century that Augustin-Jean Fresnel described the phenomenon in terms of polarization, understanding light as a wave with field components in transverse polarizations (perpendicular to the direction of the wave vector).

Explanation

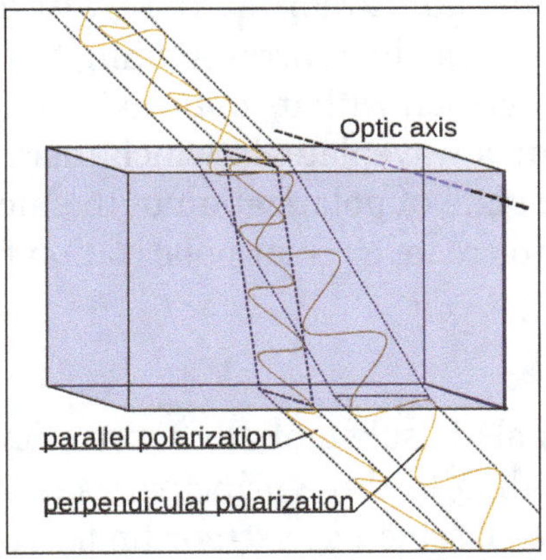

Incoming light in the parallel (p) polarization sees a different effective index of refraction than light in the perpendicular (s) polarization, and is thus refracted at a different angle.

Uniaxial Materials

The simplest type of birefringence is described as *uniaxial,* meaning that there is a single direction governing the optical anisotropy whereas all directions perpendicular to it (or at a given angle to it) are optically equivalent. Thus rotating the material around this axis does not change its optical behavior. This special direction is known as the optic axis of the material. Light propagating parallel to the optic axis (whose polarization is always *perpendicular* to the optic axis) is governed by a refractive index n_o (for "ordinary") regardless of its specific polarization. For rays with any other propagation direction, there is one linear polarization that would be perpendicular to the optic axis, and a ray with that polarization is called an *ordinary ray* and is governed by the same refractive index value n_o. However, for a ray propagating in the same direction but with a polarization perpendicular to that of the ordinary ray, the polarization direction will be partly in the direction of the optic axis, and this *extraordinary ray* will be governed by a different, *direction-dependent* refractive index. Because the index of refraction depends on the polarization, when unpolarized light enters a uniaxial birefringent material, it is split into two beams travelling in different directions, one having the polarization of the ordinary ray and the other the polarization of the extraordinary ray. The ordinary ray will always experience a refractive index of n_o, whereas the refractive index of the extraordinary ray will be in between n_o and n_e, depending on the ray direction as described by the index ellipsoid. The magnitude of the difference is quantified by the birefringence:

$$\Delta n = n_e - n_o .$$

The propagation (as well as reflection coefficient) of the ordinary ray is simply described by n_o as if there were no birefringence involved. However the extraordinary ray, as its name suggests, propagates unlike any wave in an isotropic optical material. Its refraction (and reflection) at a surface can be understood using the effective refractive index (a value in between n_o and n_e). However its power flow (given by the Poynting vector) is not exactly in the direction of the wave vector. This causes an additional shift in that beam, even when launched at normal incidence, as is popularly observed using a crystal of calcite as photographed above. Rotating the calcite crystal will cause one of the two images, that of the extraordinary ray, to rotate slightly around that of the ordinary ray, which remains fixed.

When the light propagates either along or orthogonal to the optic axis, such a lateral shift does not occur. In the first case, both polarizations are perpendicular to the optic axis and see the same effective refractive index, so there is no extraordinary ray. In the second case the extraordinary ray propagates at a different phase velocity (corresponding to n_e) but still has the power flow in the direction of the wave vector. A crystal with its optic axis in this orientation, parallel to the optical surface, may be used to create a waveplate, in which there is no distortion of the image but an intentional modification of the state of polarization of the incident wave. For instance, a quarter-wave plate is commonly used to create circular polarization from a linearly polarized source.

Biaxial Materials

The case of so-called biaxial crystals is substantially more complex. These are characterized by *three* refractive indices corresponding to three principal axes of the crystal. For most ray directions, *both* polarizations would be classified as extraordinary rays but with different effective refractive indices. Being extraordinary waves, however, the direction of power flow is not identical to the direction of the wave vector in either case.

The two refractive indices can be determined using the index ellipsoids for given directions of the polarization. Note that for biaxial crystals the index ellipsoid will *not* be an ellipsoid of revolution ("spheroid") but is described by three unequal principle refractive indices n_α, n_β and n_γ. Thus there is no axis around which a rotation leaves the optical properties invariant (as there is with uniaxial crystals whose index ellipsoid *is* a spheroid).

Although there is no axis of symmetry, there are *two* optical axes or *binormals* which are defined as directions along which light may propagate without birefringence, i.e., directions along which the wavelength is independent of polarization. For this reason, birefringent materials with three distinct refractive indices are called *biaxial*. Additionally, there are two distinct axes known as *optical ray axes* or *biradials* along which the group velocity of the light is independent of polarization.

Double Refraction

When an arbitrary beam of light strikes the surface of a birefringent material, the polarizations corresponding to the ordinary and extraordinary rays generally take somewhat different paths. Unpolarized light consists of equal amounts of energy in any two orthogonal polarizations, and even polarized light (except in special cases) will have some energy in each of these polarizations. According to Snell's law of refraction, the angle of refraction will be governed by the effective refractive index which is different between these two polarizations. This is clearly seen, for instance, in the Wollaston prism which is designed to separate incoming light into two linear polarizations using a birefringent material such as calcite.

The different angles of refraction for the two polarization components are shown in the figure at the top of the page, with the optic axis along the surface (and perpendicular to the plane of incidence), so that the angle of refraction is different for the *p* polarization (the "ordinary ray" in this case, having its polarization perpendicular to the optic axis) and the *s* polarization (the "extraordinary ray" with a polarization component along the optic axis). In addition, a distinct form of double refraction occurs in cases where the optic axis is not along the refracting surface (nor exactly normal to it); in this case the electric polarization of the birefringent material is not exactly in the direction of the wave's electric field for the extraordinary ray. The direction of power flow (given by the Poynting vector) for this inhomogenous wave is at a finite angle from the direction of the wave vector resulting in an additional separation between these beams. So even in the case of normal incidence, where the angle of refraction is zero (according to Snell's law, regardless of effective index of refraction), the energy of the extraordinary ray may be propagated at an angle. This is commonly observed using a piece of calcite cut appropriately with respect to its optic axis, placed above a paper with writing.

Much of the work involving polarization preceded the understanding of light as a transverse electromagnetic wave, and this has affected some terminology in use. Isotropic materials have symmetry in all directions and the refractive index is the same for any polarization direction. An anisotropic material is called "birefringent" because it will generally refract a single incoming ray in two directions, which we now understand correspond to the two different polarizations. This is true of either a uniaxial or biaxial material.

In a uniaxial material, one ray behaves according to the normal law of refraction (corresponding to the ordinary refractive index), so an incoming ray at normal incidence remains normal to the refracting surface. However, the other polarization can be deviated from normal incidence, which cannot be described using the law of refraction. This thus became known as the *extraordinary ray*. The terms "ordinary" and "extraordinary" are still applied to the polarization components perpendicular to and not perpendicular to the optic axis respectively, even in cases where no double refraction is involved.

A material is termed *uniaxial* when it has a single direction of symmetry in its optical behavior, which we term the optic axis. It also happens to be the axis of symmetry of the index ellipsoid (a spheroid in this case). The index ellipsoid could still be described according to the refractive indices, n_α, n_β and n_γ, along three coordinate axes, however in this case two are equal. So if $n_\alpha = n_\beta$ corresponding to the x and y axes, then the extraordinary index is n_γ corresponding to the z axis, which is also called the *optic axis* in this case.

However materials in which all three refractive indices are different are termed *biaxial* and the origin of this term is more complicated and frequently misunderstood. In a uniaxial crystal, different polarization components of a beam will travel at different phase velocities, *except* for rays in the direction of what we call the optic axis. Thus the optic axis has the particular property that rays in that direction do *not* exhibit birefringence, with all polarizations in such a beam experiencing the same index of refraction. It is very different when the three principal refractive indices are all different; then an incoming ray in any of those principle directions will still encounter two different refractive indices. But it turns out that there are two special directions (at an angle to all of the 3 axes) where the refractive indices for different polarizations are again equal. For this reason, these crystals were designated as *biaxial*, with the two "axes" in this case referring to ray directions in which propagation does not experience birefringence.

Fast and Slow Rays

In a birefringent material, a wave consists of two polarization components which generally are governed by different effective refractive indices. The so-called *slow ray* is the component for which the material has the higher effective refractive index (slower phase velocity), while the *fast ray* is the one with a lower effective refractive index. When a beam is incident on such a material from air (or any material with a lower refractive index), the slow ray is thus refracted more towards the normal than the fast ray. In the figure at the top of the page, it can be seen that refracted ray with *s* polarization in the direction of the optic axis (thus the extraordinary ray) is the slow ray in this case.

Using a thin slab of that material at normal incidence, one would implement a waveplate. In this case there is essentially no spatial separation between the polarizations, however the phase of the wave in the parallel polarization (the slow ray) will be retarded with respect to the perpendicular polarization. These directions are thus known as the slow axis and fast axis of the waveplate.

Positive or Negative

Uniaxial birefringence is classified as positive when the extraordinary index of refraction n_e is greater than the ordinary index n_o. Negative birefringence means that $\Delta n = n_e - n_o$ is less than zero. In other words, the polarization of the fast (or slow) wave is perpendicular to the optic axis when the birefringence of the crystal is positive (or negative, respectively). In the case of biaxial crystals, all three of the principal axes have different refractive indices, so this designation does not apply. But for any defined ray direction one can just as well designate the fast and slow ray polarizations.

Sources of Optical Birefringence

While birefringence is usually obtained using an anisotropic crystal, it can result from an optically isotropic material in a few ways:

- Stress birefringence results when isotropic materials are stressed or deformed (i.e., stretched or bent) causing a loss of physical isotropy and consequently a loss of isotropy in the material's permittivity tensor.

- Circular birefringence in liquids where there is an enantiomeric excess in a solution containing a molecule which has stereo isomers.
- Form birefringence, whereby structure elements such as rods, having one refractive index, are suspended in a medium with a different refractive index. When the lattice spacing is much smaller than a wavelength, such a structure is described as a metamaterial.
- By the Kerr effect, whereby an applied electric field induces birefringence at optical frequencies through the effect of nonlinear optics.
- By the Faraday effect, where a magnetic field causes some materials to become circularly birefringent (having slightly different indices of refraction for left- and right-handed circular polarizations), making the material optically active until the field is removed.
- By the self or forced alignment into thin films of amphiphilic molecules such as lipids, some surfactants or liquid crystals.

Common Birefringent Materials

The best characterized birefringent materials are crystals. Due to their specific crystal structures their refractive indices are well defined. Depending on the symmetry of a crystal structure (as determined by one of the 32 possible crystallographic point groups), crystals in that group may be forced to be isotropic (not birefringent), to have uniaxial symmetry, or neither in which case it is a biaxial crystal. The crystal structures permitting uniaxial and biaxial birefringence are noted in the two tables, below, listing the two or three principal refractive indices (at wavelength 590 nm) of some better known crystals.

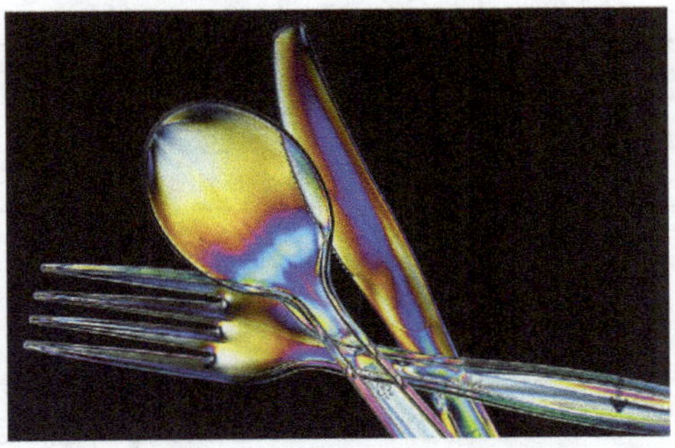

Light polarization shown on clear polystyrene cutlery between crossed polarizers.

Many plastics are birefringent because their molecules are "frozen" in a stretched conformation when the plastic is molded or extruded. For example, ordinary cellophane is birefringent. Polarizers are routinely used to detect stress in plastics such as polystyrene and polycarbonate.

Cotton fiber is birefringent because of high levels of cellulosic material in the fiber's secondary cell wall.

Polarized light microscopy is commonly used in biological tissue, as many biological materials are birefringent. Collagen, found in cartilage, tendon, bone, corneas, and several other areas in the body, is birefringent and commonly studied with polarized light microscopy. Some proteins are also birefringent, exhibiting form birefringence.

Inevitable manufacturing imperfections in optical fiber leads to birefringence, which is one cause of pulse broadening in fiber-optic communications. Such imperfections can be geometrical (lack of circular symmetry), due to stress applied to the optical fiber and due to bending of the fiber. Birefringence is *intentionally* introduced (for instance, by making the cross-section elliptical) in order to produce polarization-maintaining optical fibers.

In addition to anisotropy in the electric polarizability (electric susceptibility), anisotropy in the magnetic polarizability (magnetic permeability) can also cause birefringence. However, at optical frequencies, values of magnetic permeability for natural materials are not measurably different from μ_0, so this is not a source of optical birefringence in practice.

Uniaxial crystals, at 590 nm				
Material	Crystal system	n_o	n_e	Δn
barium borate BaB_2O_4	Trigonal	1.6776	1.5534	−0.1242
beryl $Be_3Al_2(SiO_3)_6$	Hexagonal	1.602	1.557	−0.045
calcite $CaCO_3$	Trigonal	1.658	1.486	−0.172
ice H_2O	Hexagonal	1.309	1.313	+0.004
lithium niobate $LiNbO_3$	Trigonal	2.272	2.187	−0.085
magnesium fluoride MgF_2	Tetragonal	1.380	1.385	+0.006
quartz SiO_2	Trigonal	1.544	1.553	+0.009
ruby Al_2O_3	Trigonal	1.770	1.762	−0.008
rutile TiO_2	Tetragonal	2.616	2.903	+0.287
sapphire Al_2O_3	Trigonal	1.768	1.760	−0.008
silicon carbide SiC	Hexagonal	2.647	2.693	+0.046
tourmaline (complex silicate)	Trigonal	1.669	1.638	−0.031
zircon, high $ZrSiO_4$	Tetragonal	1.960	2.015	+0.055
zircon, low $ZrSiO_4$	Tetragonal	1.920	1.967	+0.047

Biaxial crystals, at 590 nm				
Material	Crystal system	n_α	n_β	n_γ
borax $Na_2(B_4O_5)(OH)_4 \cdot 8H_2O$	Monoclinic	1.447	1.469	1.472
epsom salt $MgSO_4 \cdot 7H_2O$	Monoclinic	1.433	1.455	1.461
mica, biotite $K(Mg,Fe)_3(AlSi_3O_{10})(F,OH)_2$	Monoclinic	1.595	1.640	1.640
mica, muscovite $KAl_2(AlSi_3O_{10})(F,OH)_2$	Monoclinic	1.563	1.596	1.601
olivine $(Mg,Fe)_2SiO_4$	Orthorhombic	1.640	1.660	1.680
perovskite $CaTiO_3$	Orthorhombic	2.300	2.340	2.380
topaz $Al_2SiO_4(F,OH)_2$	Orthorhombic	1.618	1.620	1.627
ulexite $NaCaB_5O_6(OH)_6 \cdot 5H_2O$	Triclinic	1.490	1.510	1.520

Measurement

Birefringence and other polarization-based optical effects (such as optical rotation and linear or circular dichroism) can be measured by measuring the changes in the polarization of light passing through the material. These measurements are known as polarimetry. Polarized light microscopes, which contain two polarizers that are at 90° to each other on either side of the sample, are used to visualize birefringence. The addition of quarter-wave plates permit examination of circularly polarized light. Birefringence measurements have been made with phase-modulated systems for examining the transient flow behavior of fluids.

Birefringence of lipid bilayers can be measured using dual polarization interferometry. This provides a measure of the degree of order within these fluid layers and how this order is disrupted when the layer interacts with other biomolecules.

Applications

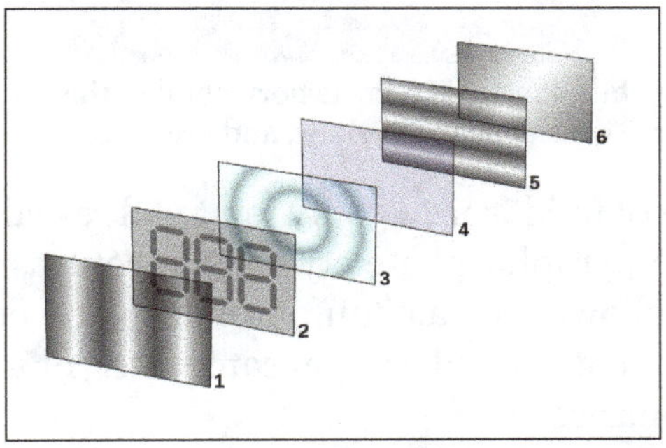

Reflective twisted-nematic liquid-crystal display.

In figure, light reflected by surface (6) (or coming from a backlight) is horizontally polarized (5) and passes through the liquid-crystal modulator (3) sandwiched in between transparent layers (2, 4) containing electrodes. Horizontally polarized light is blocked by the vertically oriented polarizer (1), except where its polarization has been rotated by the liquid crystal (3), appearing bright to the viewer.

Birefringence is used in many optical devices. Liquid-crystal displays, the most common sort of flat panel display, cause their pixels to become lighter or darker through rotation of the polarization (circular birefringence) of linearly polarized light as viewed through a sheet polarizer at the screen's surface. Similarly, light modulators modulate the intensity of light through electrically induced birefringence of polarized light followed by a polarizer. The Lyot filter is a specialized narrowband spectral filter employing the wavelength dependence of birefringence. Wave plates are thin birefringent sheets widely used in certain optical equipment for modifying the polarization state of light passing through it.

Birefringence also plays an important role in second-harmonic generation and other nonlinear optical components, as the crystals used for this purpose are almost always birefringent. By adjusting the angle of incidence, the effective refractive index of the extraordinary ray can be tuned in order to achieve phase matching, which is required for efficient operation of these devices.

Medicine

Birefringence is utilized in medical diagnostics. One powerful accessory used with optical microscopes is a pair of crossed polarizing filters. Light from the source is polarized in the x direction after passing through the first polarizer, but above the specimen is a polarizer (a so-called *analyzer*) oriented in the y direction. Therefore, no light from the source will be accepted by the analyzer, and the field will appear dark. However areas of the sample possessing birefringence will generally couple some of the x-polarized light into the y polarization; these areas will then appear bright against the dark background. Modifications to this basic principle can differentiate between positive and negative birefringence.

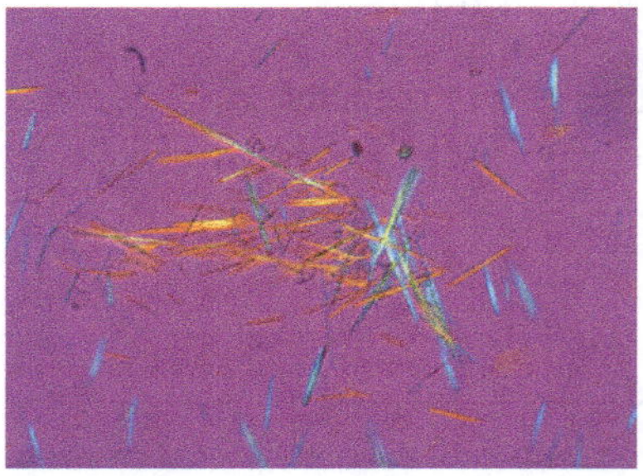

Urate crystals, with the crystals' long axis seen as horizontal in this view being parallel to that of a red compensator filter. These appear as yellow, and are thereby of negative birefringence.

For instance, needle aspiration of fluid from a gouty joint will reveal negatively birefringent monosodium urate crystals. Calcium pyrophosphate crystals, in contrast, show weak positive birefringence. Urate crystals appear yellow, and calcium pyrophosphate crystals appear blue when their long axes are aligned parallel to that of a red compensator filter, or a crystal of known birefringence is added to the sample for comparison.

Birefringence can be observed in amyloid plaques such as are found in the brains of Alzheimer's patients when stained with a dye such as Congo Red. Modified proteins such as immunoglobulin light chains abnormally accumulate between cells, forming fibrils. Multiple folds of these fibers line up and take on a beta-pleated sheet conformation. Congo red dye intercalates between the folds and, when observed under polarized light, causes birefringence.

In ophthalmology, binocular retinal birefringence screening of the *Henle fibers* (photoreceptor axons that go radially outward from the fovea) provides a reliable detection of strabismus and possibly also of anisometropic amblyopia. Furthermore, scanning laser polarimetry utilises the birefringence of the optic nerve fibre layer to indirectly quantify its thickness, which is of use in the assessment and monitoring of glaucoma.

Birefringence characteristics in sperm heads allow the selection of spermatozoa for intracytoplasmic sperm injection. Likewise, *zona imaging* uses birefringence on oocytes to select the ones with highest chances of successful pregnancy. Birefringence of particles biopsied from pulmonary nodules indicates silicosis.

Dermatologists use dermatascopes to view skin lesions. Dermatascopes use polarized light, allowing the user to view crystalline structures corresponding to dermal collagen in the skin. These structures may appear as shiny white lines or rosette shapes and are only visible under polarized dermoscopy.

Stress-induced Birefringence

Isotropic solids do not exhibit birefringence. However, when they are under mechanical stress, birefringence results. The stress can be applied externally or is "frozen in" after a birefringent plastic ware is cooled after it is manufactured using injection molding. When such a sample is placed between two crossed polarizers, colour patterns can be observed, because polarization of a light ray is rotated after passing through a birefringent material and the amount of rotation is dependent on wavelength. The experimental method called photoelasticity used for analyzing stress distribution in solids is based on the same principle. There has been recent research on using stress induced birefringence in a glass plate to generate an Optical vortex and full Poincare beams (optical beams that have every possible polarization states across its cross-section).

Color pattern of a plastic box with "frozen in" mechanical stress placed between two crossed polarizers.

Other Cases of Birefringence

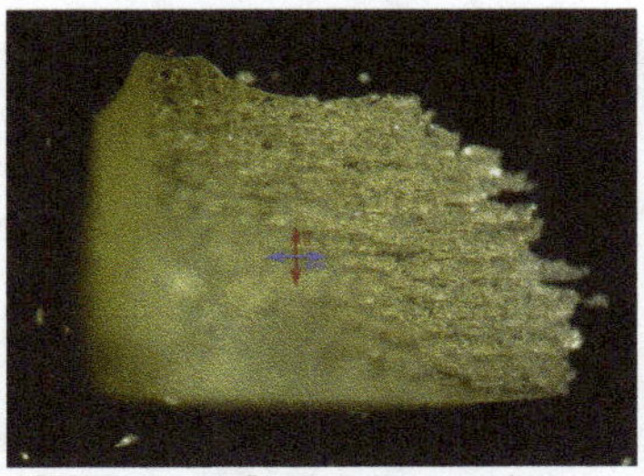

Birefringent rutile observed in different polarizations using a rotating polarizer (or analyzer).

Birefringence is observed in anisotropic elastic materials. In these materials, the two polarizations split according to their effective refractive indices, which are also sensitive to stress.

The study of birefringence in shear waves traveling through the solid Earth (the Earth's liquid core does not support shear waves) is widely used in seismology.

Birefringence is widely used in mineralogy to identify rocks, minerals, and gemstones.

Theory

In an isotropic medium (including free space) the so-called electric displacement (**D**) is just proportional to the electric field (**E**) according to **D** = ε**E** where the material's permittivity ε is just a scalar (and equal to $n^2\varepsilon_0$ where n is the index of refraction). However, in an anisotropic material exhibiting birefringence, the relationship between **D** and **E** must now be described using a tensor equation:

$$\mathbf{D} = \varepsilon\mathbf{E}$$

where ε is now a 3 × 3 permittivity tensor. We assume linearity and no magnetic permeability in the medium $\mu = \mu_0$. The electric field of a plane wave of angular frequency ω can be written in the general form:

$$\mathbf{E} = \mathbf{E}_0 e^{i(\mathbf{k}\cdot\mathbf{r} - \omega t)}$$

where r is the position vector, t is time, and $\mathbf{E}_0$ is a vector describing the electric field at $\mathbf{r}=0$, $t=0$. Then we shall find the possible wave vectors **k**. By combining Maxwell's equations for $\nabla\times\mathbf{E}$ and $\nabla\times\mathbf{H}$, we can eliminate $\mathbf{H}=\frac{1}{\mu_0}B$ to obtain:

$$-\nabla\times\nabla\times\mathbf{E}=\mu_0\frac{\partial^2}{\partial t^2}\mathbf{D}$$

With no free charges, Maxwell's equation for the divergence of D vanishes:

$$\nabla\cdot\mathbf{D}=0$$

We can apply the vector identity $\nabla\times(\nabla\times\mathbf{A})=\nabla(\nabla\cdot\mathbf{A})-\nabla^2 A$ to the left hand side of $-\nabla\times\nabla\times\mathbf{E}=\mu_0\frac{\partial^2}{\partial t^2}\mathbf{D}$, and use the spatial dependence in which each differentiation in x (for instance) results in multiplication by ik_x to find:

$$-\nabla\times\nabla\times\mathbf{E}=(\mathbf{k}\cdot\mathbf{E})\mathbf{k}-(\mathbf{k}\cdot\mathbf{k})\mathbf{E}$$

The right hand side of $-\nabla\times\nabla\times\mathbf{E}=\mu_0\frac{\partial^2}{\partial t^2}\mathbf{D}$ can be expressed in terms of **E** through application of the permittivity tensor ε and noting that differentiation in time results in multiplication by $-i\omega$, $-\nabla\times\nabla\times\mathbf{E}=\mu_0\frac{\partial^2}{\partial t^2}\mathbf{D}$ then becomes:

$$(-\mathbf{k}\cdot\mathbf{k})\mathbf{E}+(\mathbf{k}\cdot\mathbf{E})\mathbf{k}=-\mu_0\omega^2(\varepsilon\mathbf{E})$$

Applying the differentiation rule to $\nabla\cdot\mathbf{D}=0$ we find:

$$\mathbf{k}\cdot\mathbf{D}=0$$

$\mathbf{k}\cdot\mathbf{D}=0$ indicates that **D** is orthogonal to the direction of the wavevector **k**, even though that is no longer generally true for **E** as would be the case in an isotropic medium. $\mathbf{k}\cdot\mathbf{D}=0$ will not be needed for the further steps in the following derivation.

Finding the allowed values of k for a given ω is easiest done by using Cartesian coordinates with the x, y and z axes chosen in the directions of the symmetry axes of the crystal (or simply choosing z in the direction of the optic axis of a uniaxial crystal), resulting in a diagonal matrix for the permittivity tensor ε:

$$\varepsilon=\varepsilon_0\begin{bmatrix} n_x^2 & 0 & 0 \\ 0 & n_y^2 & 0 \\ 0 & 0 & n_z^2 \end{bmatrix}$$

where the diagonal values are squares of the refractive indices for polarizations along the three principal axes x, y and z. With ε in this form, and substituting in the speed of light c using $c^2 = \frac{1}{\mu_0 \varepsilon_0}$, $(-\mathbf{k}\cdot\mathbf{k})\mathbf{E} + (\mathbf{k}\cdot\mathbf{E})\mathbf{k} = -\mu_0\omega^2(\varepsilon\mathbf{E})$ becomes,

$$(-k_x^2 - k_y^2 - k_z^2)E_x + k_x^2 E_x + k_x k_y E_y + k_x k_z E_z = -\frac{\omega^2 n_x^2}{c^2} E_x$$

where E_x, E_y, E_z are the components of **E** (at any given position in space and time) and k_x, k_y, k_z are the components of k. Rearranging, we can write (and similarly for the y and z components of $(-\mathbf{k}\cdot\mathbf{k})\mathbf{E} + (\mathbf{k}\cdot\mathbf{E})\mathbf{k} = -\mu_0\omega^2(\varepsilon\mathbf{E})$),

$$\left(-k_y^2 - k_z^2 + \frac{\omega^2 n_x^2}{c^2}\right)E_x + k_x k_y E_y + k_x k_z E_z = 0$$

$$k_x k_y E_x + \left(-k_x^2 - k_z^2 + \frac{\omega^2 n_y^2}{c^2}\right)E_y + k_y k_z E_z = 0$$

$$k_x k_z E_x + k_y k_z E_y + \left(-k_x^2 - k_y^2 + \frac{\omega^2 n_z^2}{c^2}\right)E_z = 0$$

This is a set of linear equations in E_x, E_y, E_z, so it can have a nontrivial solution (that is, one other than E = 0) as long as the following determinant is zero:

$$\begin{vmatrix} \left(-k_y^2 - k_z^2 + \frac{\omega^2 n_x^2}{c^2}\right) & k_x k_y & k_x k_z \\ k_x k_y & \left(-k_x^2 - k_z^2 + \frac{\omega^2 n_y^2}{c^2}\right) & k_y k_z \\ k_x k_z & k_y k_z & \left(-k_x^2 - k_y^2 + \frac{\omega^2 n_z^2}{c^2}\right) \end{vmatrix} = 0$$

Evaluating the determinant of $\begin{vmatrix} \left(-k_y^2 - k_z^2 + \frac{\omega^2 n_x^2}{c^2}\right) & k_x k_y & k_x k_z \\ k_x k_y & \left(-k_x^2 - k_z^2 + \frac{\omega^2 n_y^2}{c^2}\right) & k_y k_z \\ k_x k_z & k_y k_z & \left(-k_x^2 - k_y^2 + \frac{\omega^2 n_z^2}{c^2}\right) \end{vmatrix} = 0$,

and rearranging the terms, we obtain:

$$\frac{\omega^4}{c^4} - \frac{\omega^2}{c^2}\left(\frac{k_x^2 + k_y^2}{n_z^2} + \frac{k_x^2 + k_z^2}{n_y^2} + \frac{k_y^2 + k_z^2}{n_x^2}\right) + \left(\frac{k_x^2}{n_y^2 n_z^2} + \frac{k_y^2}{n_x^2 n_z^2} + \frac{k_z^2}{n_x^2 n_y^2}\right)\left(k_x^2 + k_y^2 + k_z^2\right) = 0.$$

In the case of a uniaxial material, choosing the optic axis to be in the z direction so that $n_x = n_y = n_o$ and $n_z = ne$, this expression can be factored into,

$$\left(\frac{k_x^2}{n_o^2}+\frac{k_y^2}{n_o^2}+\frac{k_z^2}{n_o^2}-\frac{\omega^2}{c^2}\right)\left(\frac{k_x^2}{n_e^2}+\frac{k_y^2}{n_e^2}+\frac{k_z^2}{n_o^2}-\frac{\omega^2}{c^2}\right)=0$$

Setting either of the factors in $\left(\frac{k_x^2}{n_o^2}+\frac{k_y^2}{n_o^2}+\frac{k_z^2}{n_o^2}-\frac{\omega^2}{c^2}\right)\left(\frac{k_x^2}{n_e^2}+\frac{k_y^2}{n_e^2}+\frac{k_z^2}{n_o^2}-\frac{\omega^2}{c^2}\right)=0$ to zero will define an ellipsoidal surface in the space of wavevectors **k** that are allowed for a given ω. The first factor being zero defines a sphere; this is the solution for so-called ordinary rays, in which the effective refractive index is exactly n_o regardless of the direction of **k**. The second defines a spheroid symmetric about the z axis. This solution corresponds to the so-called extraordinary rays in which the effective refractive index is in between n_o and n_e, depending on the direction of **k**. Therefore, for any arbitrary direction of propagation (other than in the direction of the optic axis), two distinct wavevectors **k** are allowed corresponding to the polarizations of the ordinary and extraordinary rays.

For a biaxial material a similar but more complicated condition on the two waves can be described; the allowed k vectors in a specified direction now lie on one of two ellipsoids. By inspection one can see that

$$\begin{vmatrix} \left(-k_y^2-k_z^2+\frac{\omega^2 n_x^2}{c^2}\right) & k_x k_y & k_x k_z \\ k_x k_y & \left(-k_x^2-k_z^2+\frac{\omega^2 n_y^2}{c^2}\right) & k_y k_z \\ k_x k_z & k_y k_z & \left(-k_x^2-k_y^2+\frac{\omega^2 n_z^2}{c^2}\right) \end{vmatrix}=0$$

is generally satisfied for two positive values of ω. Or, for a specified optical frequency ω and direction normal to the wavefronts $\frac{k}{|\mathbf{k}|}$, it is satisfied for two wavenumbers (or propagation constants) $|\mathbf{k}|$ (and thus effective refractive indices) corresponding to the propagation of two linear polarizations in that direction.

When those two propagation constants are equal then the effective refractive index is independent of polarization, and there is consequently no birefringence encountered by a wave traveling in that particular direction. For a uniaxial crystal, this is the optic axis, the z direction according to the above construction. But when all three refractive indices (or permittivities), n_x, n_y and n_z are distinct, it can be shown that there are exactly two such directions (where the two ellipsoids intersect); these directions are not at all obvious and do not lie along any of the three principal axes (x, y, z according to the above convention). Historically that accounts for the use of the term "biaxial" for such crystals, as the existence of exactly two such special directions (considered "axes") was discovered well before polarization and birefringence were understood physically. However these two special directions are generally not of particular interest; biaxial crystals are rather specified by their three refractive indices corresponding to the three axes of symmetry.

A general state of polarization launched into the medium can always be decomposed into two waves, one in each of those two polarizations, which will then propagate with different wavenumbers $|\mathbf{k}|$. Applying the different phase of propagation to those two waves over a specified propagation distance will result in a generally *different* net polarization state at that point; this is the principle of the waveplate for instance. However, when you have a wave launched into a birefringent material at non-normal incidence, the problem is yet more complicated since the two polarization components will now not only have distinct wavenumbers but the **k** vectors will not even be in exactly the same direction. In this case the two **k** vectors are rather solutions of:

$$\begin{vmatrix} \left(-k_y^2 - k_z^2 + \frac{\omega^2 n_x^2}{c^2}\right) & k_x k_y & k_x k_z \\ k_x k_y & \left(-k_x^2 - k_z^2 + \frac{\omega^2 n_y^2}{c^2}\right) & k_y k_z \\ k_x k_z & k_y k_z & \left(-k_x^2 - k_y^2 + \frac{\omega^2 n_z^2}{c^2}\right) \end{vmatrix} = 0$$

constrained by the boundary condition which requires that the components of the two transmitted waves' k vectors, and the k vector of the incident wave, as projected onto the surface of the interface, must all be identical.

OPTICAL ENGINEERING

Optical engineering refers to the study of applications of optics. It is used for designing of optical instruments such as lenses, microscopes, telescopes, and other equipment that utilizes the properties of light. Optical engineering also includes the designing and application of optical sensors and measurement systems, lasers, fiber optic communication systems, optical disc systems. In optical engineering, instruments like the laser speckle interferometer are used to measure micro-vibrations or to measure the properties of the various masses with instruments measuring refraction. Optical engineers considers the practicality in terms of available technology, materials, costs, design methods, and also build devices which manipulate light by applying knowledge of optics.

References

- Optics, science: britannica.com, Retrieved 1 February, 2019
- Rays , modules, optics421: phys.utk.edu, Retrieved 2 March, 2019
- Reed M. Jost; Joost Felius; Eileen E. Birch (August 2014). "High sensitivity of binocular retinal birefringence screening for anisometropic amblyopia without strabismus". Journal of American Association for Pediatric Ophthalmology and Strabismus (JAAPOS). 18 (4): e5–e6. Doi:10.1016/j.jaapos.2014.07.017
- Resolution, imaging, application-notes: edmundoptics.com, Retrieved 3 April, 2019
- Illusion, topic: britannica.com, Retrieved 4 May, 2019

Optical Materials and Lens Design

CHAPTER 2

The materials which are used to construct the devices that alter or control the electromagnetic radiation in the visible, ultraviolet, and infrared spectral regions are known as optical materials. Lens designing is the designing of lens and it focuses on optical lens design, opto-mechanical system design, and optical fabrication and testing. This chapter has been carefully written to provide an easy understanding of optical materials and lens design.

OPTICAL MATERIALS

Optical materials are substances used to manipulate the flow of light. This can include reflecting, absorbing, focusing or splitting an optical beam. The efficiency of a specific material at each task is strongly wavelength dependent, thus a full understanding of the interaction between light and matter is vital.

Optical materials are fabricated into optical elements such aslenses, mirrors, windows, prisms, polarizers, detectors, and modulators. These materials serve to refract, reflect, transmit,disperse, polarize, detect, and transform light. The term "light" refers here not only to visible light but also to radiation in theadjoining ultraviolet and infrared spectral regions. At the microscopic level, atoms and their electronic configurations in thematerial interact with the electromagnetic radiation (photons) to determine the material's macroscopic optical properties suchas transmission and refraction. These optical properties are functions of the wavelength of the incident light, the temperatureof the material, the applied pressure on the material, and in certain instances the external electric and magnetic fields appliedto the material.

There is a wide range of substances that are useful as optical materials. Most optical elements are fabricated from glass,crystalline materials, polymers, or plastic materials. In the choice of a material, the most important properties are often thedegree of transparency and the refractive index, along with each property's spectral dependency. The uniformity of thematerial, the strength and hardness, temperature limits, hygroscopicity, chemical resistivity, and availability of suitablecoatings may also need to be considered.

Glass technology provided the foundation for classical optical elements, such as lenses, prisms, and filters. Glassesdeveloped for use in the visible region have internal transmittances of over 99% throughout the wavelength range of 380–780 nanometers. However, the silicate structure in glasses limits their transmission to about 2.5 micrometers in the infrared.Chalcogenide glasses, heavy-metal fluoride glasses, and heavy-metal oxide glasses extend this transmission to 8–12 μm.

Advances in the process for manufacturing optical fibers led to the present fiber-optic communication systems that operate inthe near-infrared region with windows at wavelengths of 850, 1310,

1550, and 1625 nm. An advanced fiber-optic system, LEAF (Large Effective Area Fiber), was designed to minimize nonlinearities by spreading the optical power over large areas.

The use of photolithography for printing integrated circuits has necessitated the improvement in the transmission of glassesfor the ultraviolet region. Fused silica, which transmits to about 180 nm, is well suited for the lithography in the ultravioletregion. However, the crystalline material calcium fluoride, which transmits into the ultraviolet region to about 140 nm,outperforms any glass in printing microchips using fluorine excimer lasers. Deep-ultraviolet applications of fused-silicaglasses include high-energy lasers, spacecraft windows, blanks for large astronomical mirrors, optical imaging, and cancerdetection using ultraviolet-laser-induced autofluorescence.

The need for an inexpensive, unbreakable lens that could be easily mass-produced precipitated the introduction of plasticoptics in the mid-1930s. Although the variety of plastics suitable for precision optics is limited compared to glass or crystallinematerials, plastics are often preferred when difficult or unusual shapes, lightweight elements, or economical mass-productiontechniques are required.

The softness, inhomogeneity, and susceptibility to abrasion intrinsic to plastics often restrict their application. Haze (which is the light scattering due to microscopic defects) and birefringence (resulting from stresses) are inherent to plastics. Plasticsalso exhibit large variations in the refractive index with changes in temperature. Shrinkage resulting during the processingmust be considered.

Organic synthetic polymers are emerging as key materials for information technologies. Polymers often have an advantageover inorganic materials because they can be designed and synthesized into compositions and architectures not possiblewith crystals, glasses, or plastics. They are manufactured to be durable, optically efficient, reliable, and inexpensive. Manyuses of polymers in photonic and optoelectronic devices have emerged, including light-emitting diodes, liquid-crystal–polymerphotodetectors, polymer-dispersed liquid-crystal devices (for projection television), optical-fiber amplifiers doped with organicdyes (rhodamine), organic thin-film optics, and electrooptic modulators.

Although most of the early improvements in optical devices were due to advancements in the production of glasses, thecrystalline state has taken on increasing importance. Historically, the naturally occurring crystals such as rock salt, quartz,and fluorite plus suitable detectors permitted the first extension of visible optical techniques to harness the invisible ultravioletand infrared rays. Synthetic crystal-growing techniques have made available single crystals such as lithium fluoride (of special value in the ultraviolet region, since it transmits at wavelengths down to about 120 nm), calcium fluoride, andpotassium bromide (useful as a prism at wavelengths up to about 25 μm in the infrared). Many alkali-halide crystals areimportant because they transmit into the far-infrared.

Following the invention of the transistor, germanium and silicon ushered in the use of semiconductors as infrared opticalelements or detectors. Polycrystalline forms of these semiconductors could be fabricated into windows, prisms, lenses, anddomes by casting, grinding, and polishing. Compound semiconductors such as gallium arsenide (GaAs), ternary compoundssuch as gallium aluminum arsenide ($Ga_{1-x}Al_xAs$), and quaternary compounds such as indium gallium arsenide phosphide (InGaAsP) now serve as lasers, light-emitting diodes, and photodetectors.

Single crystals are indispensable for transforming, amplifying, and modulating light. Birefringent crystals serve as retarders,or wave plates, which are used to convert the polarization state of the light. In many cases, it is desirable that the crystals notonly be birefringent, but also behave nonlinearly when exposed to very large fields such as those generated by intense laserbeams. A few examples of such non-linear crystals are ammonium dihydrogen phosphate (ADP), potassium dihydrogenphosphate (KDP), beta barium borate (BBO), lithium borate (LBO), and potassium titanyl phosphate (KTP).

Other optical materials are the liquid crystals used in displays as light valves, materials used in erasable optical disks forcomputers and in liquid cells (Kerr cells), laser dyes, dielectric multilayer films, filter materials, and the many metals (aluminum, gold, beryllium, and so forth) and alloys that are important as coating materials.

OPTICAL GLASS

Selecting a glass material is important since different glass types have different characteristics.

The index of refraction and Abbe number of a glass are typically used by designers as degrees of freedom when designing systems. The index of refraction refers to the ratio of the speed of light in a vacuum to the speed of light through a given material at a given wavelength, while the Abbe number of a material quantifies the amount of dispersion (variations in index) for a specific spectral range. For instance, a higher index of refraction generally bends light more efficiently so there is less of a need of curvature in the lens. Spherical aberration is less present in lenses with higher indices of refraction, while light travels faster through materials with lower indices of refraction. A high Abbe number generally gives less color dispersion and reduces color aberration. Also, certain glass types have different transmission wavelength regions.

The density of a glass helps determine the weight of the optical assembly and, along with lens diameter, becomes critical for weight sensitive applications. When dealing with applications involving extreme temperatures and quick temperature differentials, a glass' coefficient of expansion also becomes a key factor. Opto-mechanical designers need to keep this in mind when designing optical assemblies.

Optical Glass Properties

Today, the quality and integrity of optical glass is a fundamental assumption made by optical designers. Nearly 125 years ago, Otto Schott began a revolution by systematically researching and developing glass compositions. His development work on composition and the production process took glass manufacturing from the realm of trial and error to its state today as a truly technical material. Now optical glass properties are predictable, reproducible and homogeneous - the essential prerequisites of a technical material. The fundamental properties that characterize optical glass are refractive index, dispersion, and transmission.

Refractive Index

Refractive Index is the ratio of the speed of light in a vacuum to the speed of light in the specified material - a description of how light slows down as it passes through an optical material. The refractive index for optical glasses, n_d, is specified at a wavelength of 587.6nm (Helium d-line). Materials with a low index of refraction are commonly referred to as "crowns" whereas materials with a high index of refraction are referred to as "flints".

Dispersion

Dispersion is a description of the variation of the refractive index with wavelength. It is specified using the Abbe number, v_d, defined as $(n_d - 1)/(n_F - n_C)$ where n_F and n_C are the refractive indices at 486.1 nm (Hydrogen F-line) and 656.3 nm (Hydrogen C-line). A low Abbe number indicates high dispersion. Crown glasses tend to have lower dispersion than flints.

Transmission

Standard optical glasses offer high transmission throughout the entire visible spectrum and beyond in the near ultraviolet and near infrared ranges. Crown glasses tend to have better transmission in the NUV than do flint glasses. Flint glasses, because of their high index, feature higher Fresnel reflection loss and thus should always be specified with an anti-reflection (AR) coating.

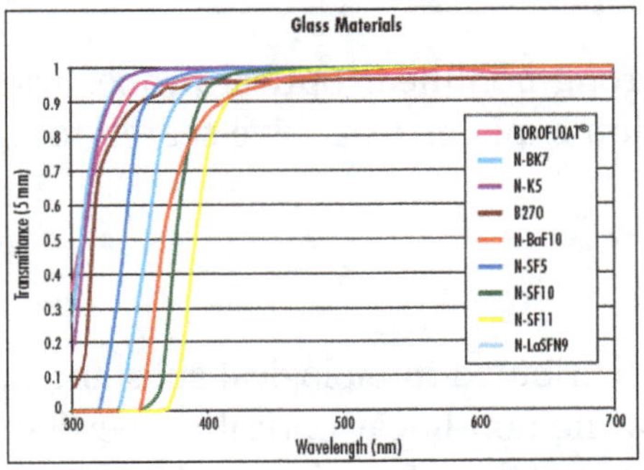

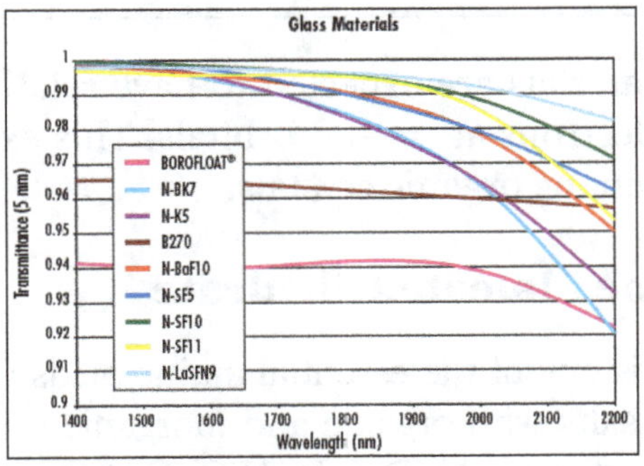

Sample Optical Glass Transmittance Curves.

Optical Glass Selection

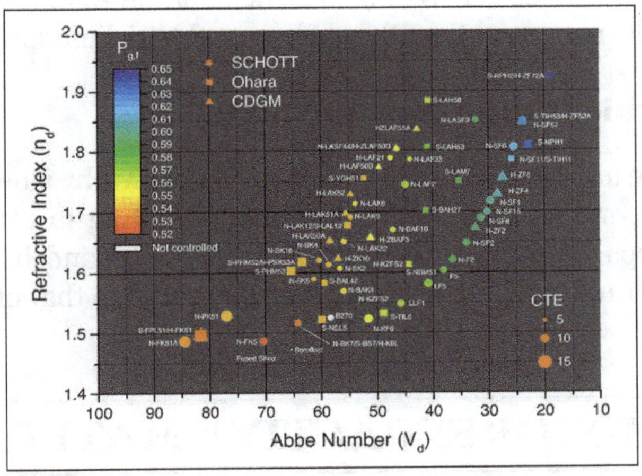

Abbe diagram with Coefficient of Thermal Expansion and Relative Partial Dispersion.

Optical systems have to be optimized for a total set of functional characteristics. Geometrical and color induced aberrations can be compensated only by the use of more than one glass type. In most cases three or more glass types are used. The requirements on optical systems for different applications cover a range so wide that they cannot be met with just a small set of glass types. So a wide range of glass types has been developed. Traditionally they are shown in the refractive index versus dispersion diagram - the Abbe diagram.

The Abbe diagram, first introduced by SCHOTT in 1923, is a long established survey of the optical glass program. Glass types are given in a two-dimensional coordinate system with the Abbe number (v_d) as x-axis and the refractive index (n_d) as y-axis. The x-axis is in reversed direction with numbers increasing to the left side.

In the Abbe diagram, glass materials are divided into type denominations like BK, SK, F, SF, etc. These "glass families" correspond to the regions in the Abbe diagram defined by the blue lines. There is a major line that separates crown glass types (last letter "K" from German "Kron" for crown) from flint glass types (last letter "F"). This line starts upwards from the bottom at Abbe number 55, steps aside at refractive index 1.60 to Abbe number 50, and continues upwards to the top.

ORGANIC NONLINEAR OPTICAL MATERIALS

Organic materials are expected to have relatively strong nonlinear optical properties due to delocalized electrons at $\pi - \pi^*$ orbitals. This expectation explains extensive search for better NLO materials among organic crystals.

L-arginine Maleate Dihydrate

L-arginine is one of the essential amino acids widely distributed in biological substances. It forms a number of salts with organic and inorganic acids showing non-linear optical properties. L-arginine maleate dihydrate (LAMD, $C_6H_{14}N_4O_2$,$C_4H_4O_4$,$2H_2O$) is one of these L-arginine salts which is a complex of strongly basic amino acid, carboxylic acid and provides useful information in relation to molecular interaction in present-day biological systems and to prebiotic self-organisms. It is also a nonlinear optical material with second harmonic generation efficiency 1.68 times that of KDP. LAMD crystals are grown from solution by solvent evaporation; they belong to the triclinic space group P1.

L-methionine L-methioninium Hydrogen Maleate

LMMM too belongs to the amino acid family. Crystals are grown by slow evaporation of an aqueous solution containing L-methionine and maleic acid, resulting in centimeter-large crystals of a non-centrosymmetric space group. They were applied for second harmonic generation of an Nd:YAG laser (wavelength 1064 nm), and SHG efficiency equal to that of KDP has been obtained.

ORGANIC PHOTOREFRACTIVE MATERIALS

Organic photorefractive materials are materials that exhibit a temporary change in refractive index when exposed to light. The changing refractive index causes light to change speed throughout the material and produce light and dark regions in the crystal. The buildup can be controlled to produce holographic images for use in biomedical scans and optical computing. The ease with which the chemical composition can be changed in organic materials makes the photorefractive effect more controllable.

Theory

There are two phenomena that, when combined together, produce the photorefractive effect. These are photoconductivity, first observed in selenium by Willoughby Smith in 1873, and the Pockels Effect, named after Friedrich Carl Alwin Pockels who studied it in 1893.

Photoconductivity is the property of a material that describes the capability of incident light of adequate wavelength to produce electric charge carriers. The Fermi level of an intrinsic semiconductor is exactly in the middle of the band gap. The densities of free electrons *n* in the conduction band and free holes *h* in the valence band can be found through equations:

$$n = N_c \frac{e^{-(E_c - E_F)}}{k_B T}$$

and

$$h = N_v \frac{e^{-(E_c - E_F)}}{k_B T}$$

where N_C and N_V are the densities of states at the bottom of the conduction band and the top of the valence band, respectively, E_C and E_V are the corresponding energies, E_F is the Fermi level, k_B is Boltzmann's constant and T is the absolute temperature. Addition of impurities into the semiconductor, or doping, produces excess holes or electrons, which, with sufficient density, may pin the Fermi level to the impurities' position.

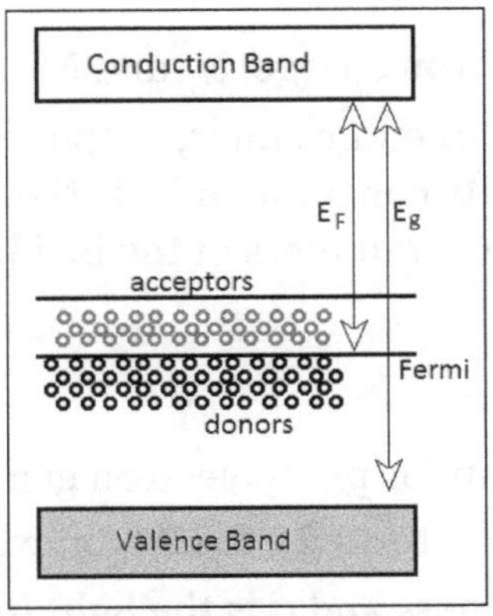

Doped semiconductor: Fermi levels pinned at the position of the dopant in the band gap.

A sufficiently energetic light can excite charge carriers so much that they will populate the initially empty localized levels. Then, the density of free carriers in the conduction and the valence band will increase. To account for these changes, steady-state Fermi levels are defined for electrons to be E_{Fn} and, for holes $-E_{Fp}$. The densities n and h are, then equal to:

$$n = N_c \frac{e^{-(E_c - E_{Fn})}}{k_B T}$$

$$h = N_v \frac{e^{-(E_{Fp} - E_v)}}{k_B T}$$

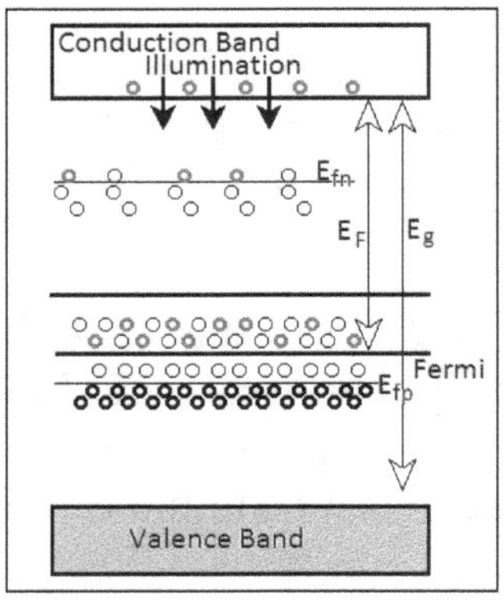

Doped semiconductor: Fermi level E_f and quasi-Fermi levels upon illumination.

The localized states between E_{Fn} and E_{Fp} are known as 'photoactive centers'. The charge carriers remain in these states for a long time until they recombine with an oppositely charged carrier. The states outside the $E_{Fn} - E_{Fp}$ energy, however, relax their charge carriers to the nearest extended states.

The effect of incident light on the conductivity of the material depends on the energy of light and material. Differently-doped materials may have several different types of photoactive centers, each of which requires a different mathematical treatment. However, it is not very difficult to show the relationship between incident light and conductivity in a material with only one type of charge carrier and one type of a photoactive center. The dark conductivity of such a material is given by,

$$\sigma_d = e\left(N_d - N_d^+\right)\beta\mu\tau$$

where σ_d is the conductivity, e = electron charge, N_D and N_D^+, are the densities of total photoactive centers and ionized empty electron acceptor states, respectively, β is the thermal photoelectron generation coefficient, μ is the mobility constant and τ is the photoelectron lifetime. The equation for photoconductivity substitutes the parameters of the incident light for β and is,

$$\sigma_{ph} = e\left(N_d - N_d^+\right)\frac{sI\mu\tau}{h\nu}$$

in which s is the effective cross-section for photoelectron generation, h is the Planck constant, ν is the frequency of incident light, and the term $I = I_0 e^{-\alpha z}$ in which I_0 is the incident irradiance, z is the coordinate along the crystal thickness and α is the light intensity loss coefficient.

The electro-optic effect is a change of the optical properties of a given material in response to an electric field. There are many different occurrences, all of which are in the subgroup of the electro-optic effect, and Pockels effect is one of these occurrences. Essentially, the Pockels effect is the change of the material's refractive index induced by an applied electric field. The refractive index of a material is the factor by which the phase velocity is decreased relative to the velocity of light in vacuum. At a microscale, such a decrease occurs because of a disturbance in the charges of each atom after being subjected to the electromagnetic field of the incident light. As the electrons move around energy levels, some energy is released as an electromagnetic wave at the same frequency but with a phase delay. The apparent light in a medium is a superposition of all of the waves released in such way, and so the resulting light wave has shorter wavelength but the same frequency and the light wave's phase speed is slowed down.

Whether or not the material will exhibit Pockels effect depends on its symmetry. Both centrosymmetric and non-centrosymmetric media will exhibit an effect similar to Pockels, the Kerr effect. The refractive index change will be proportional to the square of the electric field strength and will therefore be much weaker than the Pockels effect. It is only the non-centrosymmetric materials that can exhibit the Pockels effect: for instance, lithium tantalite (trigonal crystal) or gallium arsenide (zinc-blende crystal); as well as poled polymers with specifically designed organic molecules.

It is possible to describe the Pockels effect mathematically by first introducing the index ellipsoid – a concept relating the orientation and relative magnitude of the material's refractive indices. The ellipsoid is defined by,

$$\frac{R_1^2}{\varepsilon_1}+\frac{R_2^2}{\varepsilon_2}+\frac{R_3^2}{\varepsilon_3}=1$$

in which ε_i is the relative permittivity along the x, y, or z axis, and R is the reduced displacement vector defined as $D_i / \sqrt{8\pi W}$ in which D_i is the electric displacement vector and W is the field energy. The electric field will induce a deformation in R_i as according to:

$$\Delta R_i^2 = \sum_{j=1}^{3} r_{ij} E_j$$

in which E is the applied electric field, and r_{ij} is a coefficient that depends on the crystal symmetry and the orientation of the coordinate system with respect to the crystal axes. Some of these coefficients will usually be equal to zero.

Organic Photorefractive Materials

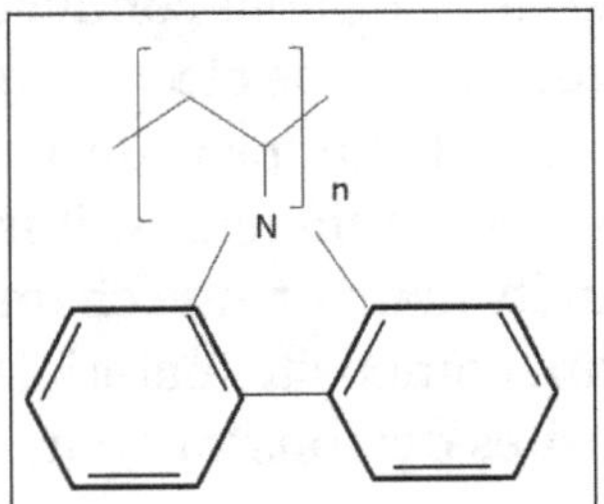

Chemical structure of PVK.

In general, photorefractive materials can be classified into the following categories, the border between categories may not be sharp in each case:

- Inorganic crystal and compound semiconductor.
- Multiple quantum well structures.
- Organic crystalline materials.
- Polymer dispersed Liquid crystalline materials (PDLC).
- Organic amorphous materials.

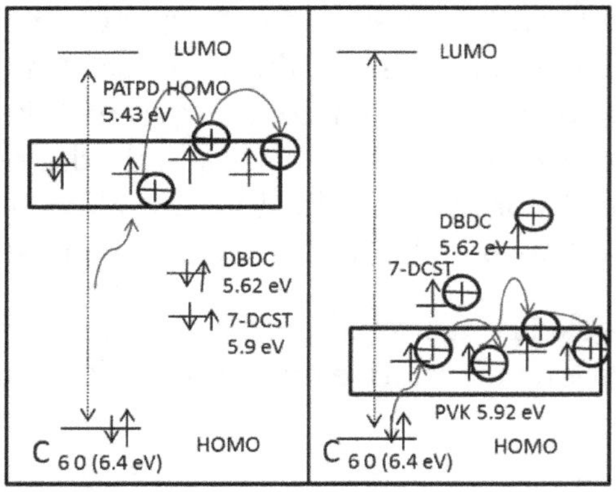

Electronic states of PATPD (a) and PVK (b)-based composite with DBDC and 7-DCST as chromophores and C60 as sensitizer.

In the field of this research, initial investigations were mainly carried out with inorganic semiconductors. There have been huge varieties of inorganic crystals such as $BaTiO_3$, $KNbO_3$, $LiNbO_3$ and inorganic compound semiconductors such as GaAs, InP, CdTe are reported in literature. First photorefractive (PR) effect in organic materials was reported in 1991 and then, research of organic photorefractive materials has drawn major attention in recent years compare to inorganic PR semiconductors. This is due to mainly cost effectiveness, relatively easy synthetic procedure, and tunable properties through modifications of chemical or compositional changes.

Polymer or polymer composite materials have shown excellent photorefractive properties of 100% diffraction efficiency. Most recently, amorphous composites of low glass transition temperature have emerged as highly efficient PR materials. These two classes of organic PR materials are also mostly investigated field. These composite materials have four components -conducting materials, sensitizer, chromophore, and other dopant molecules to be discussed in terms of PR effect. According to the literature, design strategy of hole conductors is mainly p-type based and the issues on the sensitizing are accentuated on n-type electron accepting materials, which are usually of very low content in the blends and thus do not provide a complementary path for electron conduction. In recent publications on organic PR materials, it is common to incorporate a polymeric material with charge transport units in its main or side chain. In this way, the polymer also serves as a host matrix to provide the resultant composite material with a sufficient viscosity for reasons of processing. Most guest-host composites demonstrated in the literature so far are based on hole conducting polymeric materials.

The vast majority of the polymers are based on carbazole containing polymers like poly-(N-vinyl carbazole) (PVK) and polysiloxanes (PSX). PVK is the well studied system for huge varieties of applications. In polymers, charge is transported through the HOMO and the mobility is influenced by the nature of the dopant mixed into the polymer, also it depends on the amount of dopant which may exceed 50 weight percent of the composite for guest-host materials. The mobility decreases as the concentration of charge-transport moieties decreases, and the dopant's polarity and concentration increases.

Besides the mobility, the ionization potential of the polymer and the respective dopant has also significant importance. The relative position of the polymer HOMO with respect to the ionization potential of the other components of the blends determines the extent of extrinsic hole traps in the material. TPD (tetraphenyldiaminophenyl) based materials are known to exhibit higher charge carrier mobilities and lower ionization potentials compare to carbazole based (PVK) materials. The low ionization potentials of the TPD based materials greatly enhance the photoconductivity of the materials. This is partly due to the enhanced complexation of the hole conductor, which is an electron donor, with the sensitizing agents, which is an electron acceptor. It was reported a dramatic increase of the photogeneration efficiency from 0.3% to 100% by lowering the ionization potential from 5.90 eV (PVK) to 5.39 eV (TPD derivative PATPD). This is schematically explained in the diagram using the electronic states of PVK and PATPD.

Applications

As of 2011, no commercial products utilizing organic photorefractive materials exist. All applications described are speculative or performed in research laboratories. Large DC fields required to produce holograms lead to dielectric breakdown not suitable outside the laboratory.

Reusable Holographic Displays

Many materials exist for recording static, permanent holograms including photopolymers, silver halide films, photoresists, dichromated gelatin, and photorefractives. Materials vary in their maximum diffraction efficiency, required power consumption, and resolution. Photorefractives have a high diffraction efficiency, an average-low power consumption, and a high resolution.

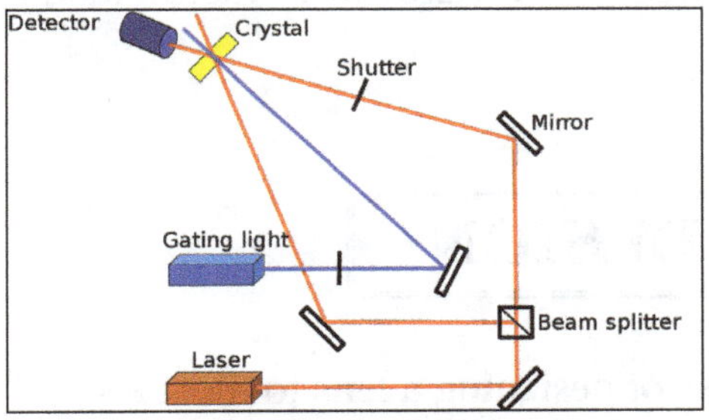

Recording of Holographic Image.

Updatable holograms that do not require glasses are attractive for medical and military imaging. The materials properties required to produce updatable holograms are 100% diffraction efficiency, fast writing time, long image persistence, fast erasing time, and large area. Inorganic materials capable of rapid updating exist but are difficult to grow larger than a cubic centimeter. Liquid crystal 3D displays exist but require complex computation to produce images which limits their refresh rate and size.

Blanche et al. demonstrated in 2008 a 4 in. x 4 in. display that refreshed every few minutes and lasted several hours. Organic photorefractive materials are capable of kHz refresh rates though it is limited by material sensitivity and laser power. Material sensitivity demonstrated in 2010 require kW pulsed lasers.

Tunable Color Filter

White light passed through an organic photorefractive diffraction grating, leads to the absorption of wavelengths generated by surface plasmon resonance and the reflection of complementary wavelengths. The period of the diffraction grating may be adjusted by modifying to control the wavelengths of the reflected light. This could be used for filter channels, optical attenuators, and optical color filters.

Optical Communications

Free-space optical communications (FSO) can be used for high-bandwidth communication of data by utilizing high frequency lasers. Phase distortions created by the atmosphere can be corrected by a four-wave mixing process utilizing organic photorefractive holograms. The nature of FSO allows images to be transmitted at near original quality in real-time. The correction also corrects for moving images.

Image and Signal Processing

Organic photorefractive materials are a nonlinear medium in which large amounts of information can be recorded and read. Holograms due to the inherent parallel nature of optical recording are able to quickly process large amounts of data. Holograms that can be quickly produced and read can be used to verify the authenticity of documents similar to a watermark Organic photorefractive correlators use matched filter and Joint Fourier Transform configurations.

Logical functions (AND, OR, NOR, XOR, NOT) were carried out using two-wave signal processing. High diffraction efficiency allowed a CCD detector to distinguish between light pixels (1 bits) and dark pixels (0 bits).

OPTICAL LENS DESIGN

Optical lens design is the process of designing a lens to meet a set of performance requirements and constraints, including cost and manufacturing limitations. Parameters include surface profile types (spherical, aspheric, holographic, diffractive, etc.), as well as radius of curvature, distance to the next surface, material type and optionally tilt and decenter. The process is computationally intensive, using ray tracing or other techniques to model how the lens affects light that passes through it.

Design Requirements

Performance requirements can include:

- Optical performance (image quality): This is quantified by various metrics, including encircled energy, modulation transfer function, Strehl ratio, ghost reflection control, and pupil performance (size, location and aberration control); the choice of the image quality metric is application specific.
- Physical requirements such as weight, static volume, dynamic volume, center of gravity and overall configuration requirements.
- Environmental requirements: ranges for temperature, pressure, vibration and electromagnetic shielding.

Design constraints can include realistic lens element center and edge thicknesses, minimum and maximum air-spaces between lenses, maximum constraints on entrance and exit angles, physically realizable glass index of refraction and dispersion properties.

Manufacturing costs and delivery schedules are also a major part of optical design. The price of an optical glass blank of given dimensions can vary by a factor of fifty or more, depending on the size, glass type, index homogeneity quality, and availability, with BK7 usually being the cheapest. Costs for larger and thicker optical blanks of a given material, above 100–150 mm, usually increase faster than the physical volume due to increased blank annealing time required to achieve acceptable index homogeneity and internal stress birefringence levels throughout the blank volume. Availability of glass blanks is driven by how frequently a particular glass type is made by a given manufacturer, and can seriously affect manufacturing cost and schedule.

Process

Lenses can first be designed using paraxial theory to position images and pupils, then real surfaces inserted and optimized. Paraxial theory can be skipped in simpler cases and the lens directly optimized using real surfaces. Lenses are first designed using average index of refraction and dispersion properties published in the glass manufacturer's catalog and though glass model calculations. However, the properties of the real glass blanks will vary from this ideal; index of refraction values can vary by as much as 0.0003 or more from catalog values, and dispersion can vary slightly. These changes in index and dispersion can sometimes be enough to affect the lens focus location and imaging performance in highly corrected systems.

The lens blank manufacturing process is as follows:

- The glass batch ingredients for a desired glass type are mixed in a powder state.
- The powder mixture is melted together in a furnace.
- The fluid is further mixed while molten to maximize batch homogeneity.
- Poured into lens blanks.
- Annealed according to empirically determined time-temperature schedules.

The glass blank pedigree, or "melt data", can be determined for a given glass batch by making small precision prisms from various locations in the batch and measuring their index of refraction on a spectrometer, typically at five or more wavelengths. Lens design programs have curve fitting routines that can fit the melt data to a selected dispersion curve, from which the index of refraction at any wavelength within the fitted wavelength range can be calculated. A re-optimization, or "melt re-comp", can then be performed on the lens design using measured index of refraction data where available. When manufactured, the resulting lens performance will more closely match the desired requirements than if average glass catalog values for index of refraction were assumed.

Delivery schedules are impacted by glass and mirror blank availability and lead times to acquire, the amount of tooling a shop must fabricate prior to starting on a project, the manufacturing tolerances on the parts (tighter tolerances mean longer fab times), the complexity of any optical coatings that must be applied to the finished parts, further complexities in mounting or bonding lens elements into cells and in the overall lens system assembly, and any post-assembly alignment and quality control testing and tooling required. Tooling costs and delivery schedules can be reduced by using existing tooling at any given shop wherever possible, and by maximizing manufacturing tolerances to the extent possible.

Lens Optimization

A simple two-element air-spaced lens has nine variables (four radii of curvature, two thicknesses, one airspace thickness, and two glass types). A multi-configuration lens corrected over a wide spectral band and field of view over a range of focal lengths and over a realistic temperature range can have a complex design volume having over one hundred dimensions.

Lens optimization techniques that can navigate this multi-dimensional space and proceed to local minima have been studied since the 1940s, beginning with early work by James G. Baker, and later by Feder, Wynne, Glatzel, Grey and others. Prior to the development of digital computers, lens optimization was a hand-calculation task using trigonometric and logarithmic tables to plot 2-D cuts through the multi-dimensional space. Computerized ray tracing allows the performance of a lens to be modelled quickly, so that the design space can be searched rapidly. This allows design concepts to be rapidly refined. In most cases the designer must first choose a viable design for the optical system, and then numerical modelling is used to refine it. The designer ensures that designs optimized by the computer meet all requirements, and makes adjustments or restarts the process when they do not.

OPTO-MECHANICAL SYSTEM DESIGN

Optical elements have been developed due to development of manufacturing tools, and we can use various kinds of mirrors, lenses, prisms or filters in various systems, and we can get high-quality elements. When we develop optical systems, we need to decide specifications according to cost, size or image quality.

Figure is an example of imaging microscope optical system. This includes features which we have to decide when we are designing basic optical systems.

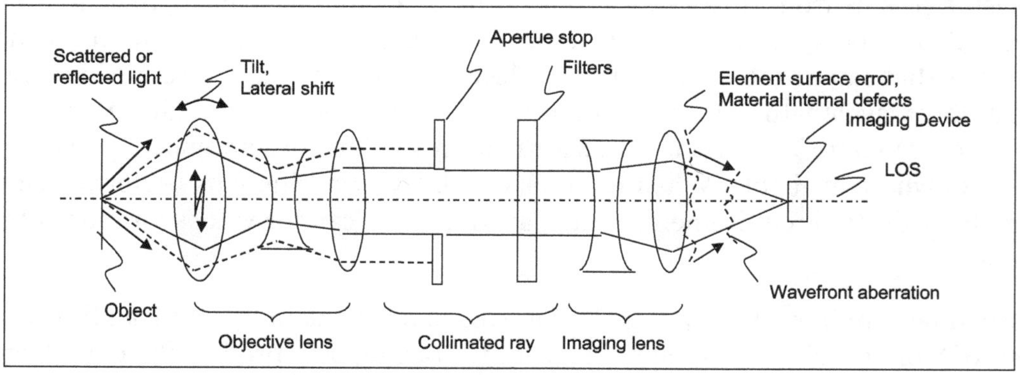

An example of imaging microscope optical system.

Specifications of Optical System with Imaging Devices

When we design optical systems, we need to decide specifications of optical systems. We need to decide following basic specifications, (1) Magnification and focal length, (2) Resolution and depth of focus.

Magnification and Focal Length

We should decide magnification of designed system according to the resolution of the image. We should decide it at first. Magnification for fundamental optical system is shown in figure. There are 2 lenses and object with height of ho. Object is placed at front focal plane of an objective lens, and an image is given at back focal plane of an imaging lens. Magnification of this optical system m is given by the following equation. This shows that we can calculate back focal length *fi* after we decide front focal length *fi* and magnification *m*.

The distance between the objective lens and imaging lens need not to be $f_o + f_i$ because the ray is collimated between the 2 lenses. The image height is independent of this distance.

$$m = \frac{h_i}{h_0} = \frac{f_i}{f_0} \rightarrow f_i = m \cdot f_0$$

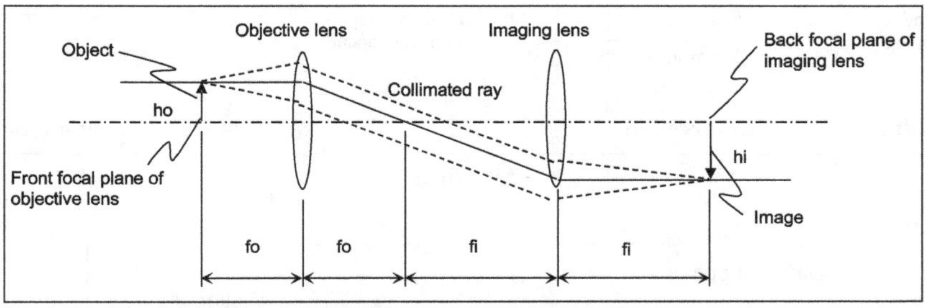

The fundamental optical system with 2 lenses.

Resolution and Depth of Focus

Resolution and depth of focus (DoF) depend on system F#. F# is important factor for image brightness and resolution of optical system, which is defined by following equation. We can get higher resolution with larger F#, but the depth of focus (DoF) becomes shorter. We have to balance both the DoF and resolution. There are 2 pictures taken with different $F\#$ in figure. The background is focused in The picture because the camera with large $F\#$ has long DoF. On the other hand, the background is defocused in the picture because the camera with small F# has short DoF. We use aperture stop in order to control $F\#$.

$$F\# = \frac{f}{D}$$

Where f is the focal length of optical element, and D is the diameter. Resolution is also decided with F#. We can get high resolution with small F#. The resolution is described with Airy disc, explained in figure. Airy disc is the size of image of a small point. The size of Airy disc is given by following equation. The resolution of optical system is often described by the Airy disc.

$$d = 2.44F \cdot \lambda .$$

Where d is diameter of Airy disc. Resolution is a minimum distance in which two points can be distinguished, which is explained in figure Resolution is almost the same size as the size of the Airy disc.

Picture with large F#(left, f5.6) and with large F#(right, f/32).

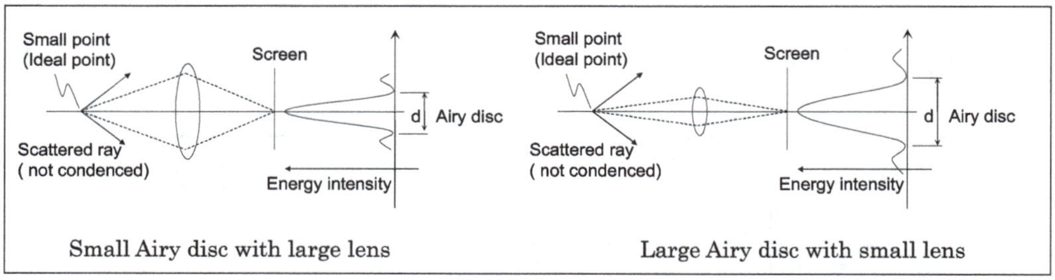

Airy disc.

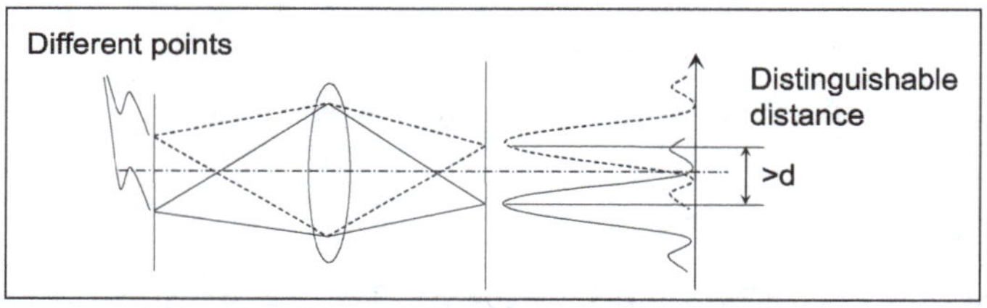

Resolution of optics and Airy disc.

Optical Elements

After we decide optical specifications, we need to think of what optical elements to use.

Lens, Mirrors and Prisms

Lens, mirrors and prisms are most basic optical elements. These deviate rays and we can condense rays or rotate images. At first we decide what lens to use and calculate focal length or other basic features.

Image Sensor

Image sensor transfers light energy to electrical energy. There are various kinds of sensors such as CMOS and CCD sensors, and it provides with each merit. CCD or CMOS sensor is used for imaging devices such as digital camera or digital photo copier. PMT is a very high sensitivity photo detector which is used for detection few photon. Table shows merit and demerit of each sensor. We usually use CCD area sensors for imaging systems. CCD, CMOS image sensor and PMT is shown in figure. CCD and CMOS is almost the same size, but PMT is much larger.

Table: Image sensor type.

	Merit	Demerit
CCD (charge coupled device).	High sensitivity, low noise.	High power consumption, high driving voltage, various power supply, difficult to manufacture.
CMOS (complementary metal oxide semiconductor).	Low power consumption, easy to manufacture, no smear.	Low sensitivity, fixed patteren noise.
PMT (photo multiplier).	Very high sensitivity.	Larger than CCD or CMOS.

(a) CCD (b) CMOS (C) PMT

Image sensors.

Filters

Filters are used to attenuate specific wavelength of rays. We should use filters. when we want to control brightness or colors of images, or need to detect specific wavelength.

There are various kinds of filters such as band-pass filter that attenuate excluding specific wavelength rays. ND filter attenuates thoroughly over a wide wavelength band, so we can control amount of rays without changing system F#. Sharp cutting filters that cut outside inferred ray or outside. Because of these features, these filters look like colored glass window. Light transmission spectrum of each filter is shown in figure.

PL (polarized light) filter can pass a specific polarized light, which is used to suppress reflected light and you can see them used for sunglasses or camera filter which can suppress reflected light from surface of water or other reflective surfaces.

We can find many other filters from optical components manufacturer website and need to choose appropriate filters according to our purpose.

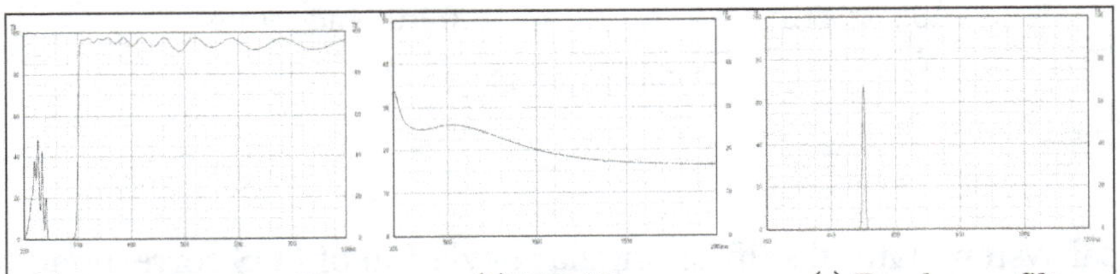

(a) Sharp-cutting filter (b) ND filter (c) Band-pass filter

Spectrum of band-pass filter(for 500nm).

Aperture Stop

Aperture stop is used in optical systems in order to change system F# to limit light amount to control the brightness of the images, DoF and resolution. Aperture stop is shown in figure. Aperture stop is placed to limit amount of light most effectively.

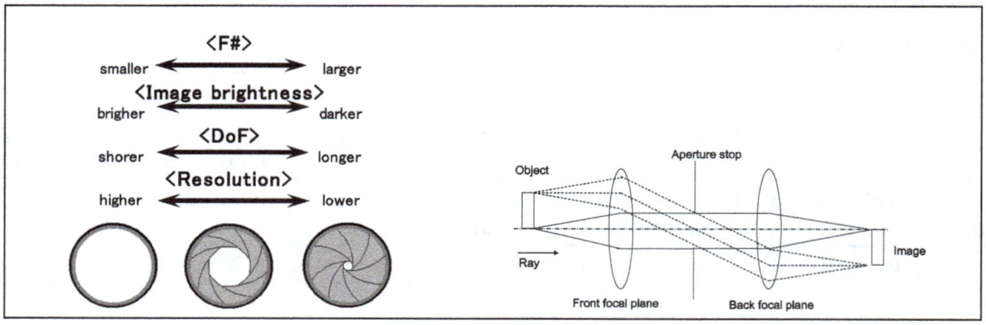

Aperture stop.

Adjustments and Mounts

We need to consider adjusting optical components with precision. It is necessary to balance precision and cost. There are various kinds of kinematic mounts. It is preferable to decide it in consideration of the method of the mount and the processing cost. Figure shows kinematic mirror holders. Kinematic mounts are often used for adjustment of element tilt. These are used for adjustment of mirror itself, but it is difficult for small lens system like a camera. Therefore it is desirable to loosen tilt error.

Figure shows the structure of kinematic mount for 2-axis tilt. The ball constrains the motion of the optical element. Figure shows the ball constraints. These provide translation constraint, XYZ motion, but rotational motion is free. Kinematic mounts realize tilt or other necessary motion with these constraint structures.

(a)2 axis tilt (b)tilt and rotate

Kinematic holder.

Line of Sight

In design of optical system, lateral shift or angular deviation of LOS correspond to the image motion. It is necessary to minimize the image motion and optimize element position error. We need to think of 6 factor of element motion, which is shown in table.

Table: System Los change with optical element motion.

	Motion type	Los angle	Image shift
Lens motion	(1) Tilt 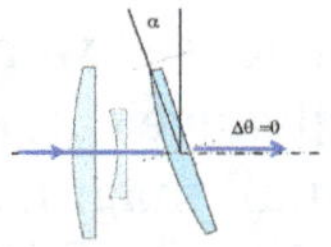	$\Delta\theta \approx 0$	$\Delta y \approx 0$
	(2) Lateral shift 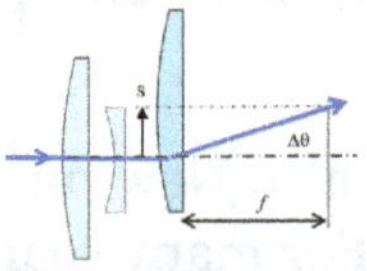	$\Delta\theta \approx \frac{s}{f}$	$\Delta y \approx \Delta\theta.d$ (d is distance between image plane and back principle point)
Mirror motion	(3) Tilt 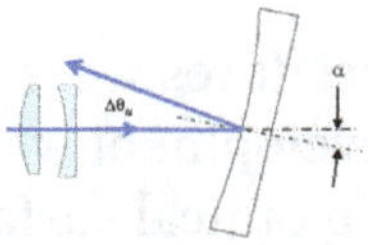	$\Delta\theta \approx 2\alpha$	$\Delta y \approx \Delta\theta.d$ (d is distance between image plane and back principle point)
	(4) Lateral shift 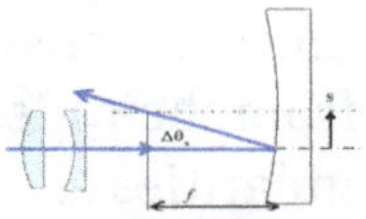	$\Delta\theta \approx \frac{s}{f}$	$\Delta y \approx \Delta\theta.d$ (d is distance between image plane and back principle point)
Filter motion	(5) Tilt 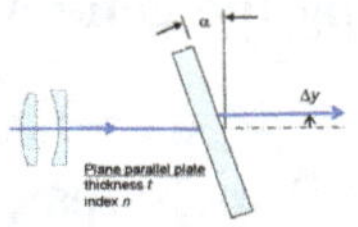	$\Delta\theta \approx 0$	$\Delta y \approx \frac{\alpha t(n-1)}{n}$
	(6) Lateral shift 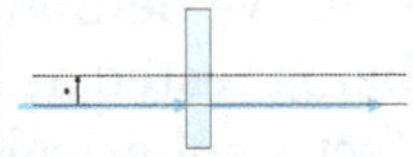	$\Delta\theta \approx 0$ (no effect)	$\Delta y \approx 0$ (no effect)

Wavefront Aberration

At first, we design optical system ignoring elements surface error. In practice, there are defects or distortion on the surfaces of optical elements, which effect to wavefront error. Wavefront error cause degradation of optical resolution.

Wavefront error is important in high-resolution system, but it is not so important if we treat low resolution optics. We need to decide quality of surface error according to resolution and cost. We need to do following three tolerance analysis. The total wavefront error is given by RSS (root sum square) of each element effect.

1) Assembly torrelance,
2) Lenses quality tolerance,
3) Thermal changes.

OPTICAL FABRICATION AND TESTING

Optical fabrication and testing techniques are more than two thousand years old. It was discovered in 1927 that as far back as 1600 B.C., crystal magnifier lenses were made in Crete. Spectacles were invented around the year 1260. The next great advance came when the Dutch microscopist Anthony von Leewenhoek in 1719 made his own small magnifying lenses to examine microscopic objects. He discovered microorganisms with his small lenses. The manufacturing details until this time are not well known, and probably no attempt was made to test the quality of the surfaces.

The traditional method of polishing with pitch was already known to Newton. He used pitch to polish his reflecting telescope, but very likely he was not the first. For many years, the testing and fabricating methods remained the same, with small variations.

In parallel with the progress in fabrication techniques, testing procedures were also generated to determine the quality of the lenses. Probably the greatest testing development was that of Foucault in France, when he invented the knife edge test and the quality of an optical surface was evaluated for the first time.

Foucault's test may be considered a method for detecting the presence of transverse aberrations. This is accomplished by intercepting with a knife the rays deviated from their ideal trajectory. This test is extremely sensitive. If the wavefront is nearly spherical, irregularities as small as a fraction of the wavelength of the light may be easily detected. This is one of the simplest and most powerful qualitative tests for observing small irregularities and evaluating the general smoothness of a concave spherical surface.

Another early important development in the testing field was the Ronchi test, invented by Vasco Ronchi in Italy in 1923. A coarse ruling is placed in the convergent light beam coming out from the surface under test, near its focus. The observer places its eye behind the ruling. The dark bands in the ruling are then projected back to the surface being tested as shadows. These shadows will be straight and parallel only if the wavefront is perfectly spherical. Otherwise, the fringes will be curves whose shape and separation will depend on the wavefront deformations. The main virtue of this test is that each type of aberration wavefront produces a characteristic Ronchi pattern. Thus, the aberrations in the optical system may be easily identified, and their magnitude estimated.

The Importance of Testing

Most of the time, comtemporary optical surfaces have a flat or a spherical shape, but they may also be toroidal or rotationally symmetric aspheric. It is easy to understand that an aspherical surface is more difficult to manufacture. In this process, the most difficult step is its testing. An aspherical surface can only be made as good as it can be tested. Hence, it is important to develop good testing methods.

Probably the first quantitative test useful for aspherical surfaces was invented in Germany by Hartmann. It is a method for determining the figure of a concave aspherical mirror. In this test, a point light source illuminates the optical surface, covered with a screen with a rectangular array of holes in front of it. Then, the position of the beams reflected through each hole on the screen is measured in a plane near the center of curvature to determine the value of the transverse aberration at each point.

The next step was the invention of the interferometric tests, mainly the Fizeau and Twyman-Green interferometers. Later, several other types of interferometers—such as the lateral shear, radial shear, and common path interferometers—were developed. Since about 1969, the use of the laser as a light source in interferometry has greatly simplified interferometric optical systems. Unfortunately, for many years these tests remained primarily qualitative, since quantitative evaluation required taking a photographic image performing extensive manual calculations.

A null interferometric test is one that produces either straight or no fringes when the optical surface is perfect. Thus a null test is simple to interpret. Frequently, a null test can not be obtained without some additional optical elements that compensate the spherical aberration of the optical element under test if it is aspheric. Many different types of compensators have been invented. Some are reflective elements, or lenses like the Offner compensator. Others are diffractive elements like real or computer generated holograms.

Enter the Computer

The advent of computers and electro-optical detectors in the early '60s allowed faster and simpler quantitative interferometric evaluations. The common method for evaluating an interferogram consists of two main steps. The first is the measuring of some samples for the fringe positions. These measurements are made in many ways, for example, by means of a measuring microscope, a digitizing tablet, or a video camera connected to a computer. If the wavefront is very irregular or the test is not null, the data points will not be uniformly distributed over the wavefront. Thus, information about many large zones is lost.

The second step is the evaluation of the entire wavefront, interpolating intermediate values by any of several existing methods. One method is to fit the wavefront data to a twodimensional polynomial by means of a least squares fitting, or by the use of splines, typically by means of a polynomial fitting to the measured data.

Phase Shifting

The problems of traditional interferometric methods have been overcome by the phase shifting interferometric techniques, where the density of sampled data points as well as the sensitivity and accuracy of the test is constant over the wavefront.

In addition, phase shifting interferometry is simple and fast, thanks to modern tools like the array detectors (CAD) and the microprocessors. Most of the conventional interferometers—Fizeau, Twyman-Green, etc.—have been adapted to perform the phase shifting techniques, the current trend in interferometry.

Phase shifting interferometric techniques began less than 20 years ago, with Crane, Moore, Bruning et al., and many others. In phase shifting interferometers, the reference wavefront is moved along the direction of propagation, with respect to the wavefront under test, changing in this manner their phase difference. The interference fringes then change their position, hence the initial name "fringe scanning" for these techniques.

By measuring the irradiance changes for different phase shifts, it is possible to determine the difference in phase between the wavefront under test and the reference wavefront, for that measured point over the wavefront. By obtaining this phase difference for many points over the wavefront, the complete wavefront shape is thus determined.

If we consider any fixed point in the interferogram, the phase difference between the two wavefronts has to be changed to make several measurements. This change in the phase may be accomplished only if the frequency of one of the beams is modified to form beats. This may be done in a continuous fashion using certain devices, but only for a relatively short period of time with some other devices. Since the frequency can be modified in a permanent way, this technique is frequently called "ACinterferometry," "heterodyne interferometry," or a "frequency shift interferometer." However, the name "phase shifting interferomtry" is more common.

In phase shifting interferometry, each detector must have a phase difference smaller than π with the closest neighboring detector to avoid 2π phase ambiguities and ensure phase continuity. In other words, there should be at least two detector elements for each fringe. If the slope of the wavefront is very large, the fringes will be too close together and the number of detector elements would be extremely large. A solution to this problem is to use two different wavelengths— λ_1 and λ_2 —simultaneously. The group wavelength or equivalent wavelength λ_{eq} is much longer than any of the two components. This clever idea was proposed by Yaou-Yan Chang and J.C. Wyant.

The new electronic phase shifting digital techniques have several distinct advantages, among which we may mention: a) improved accuracy of the interferometric measurements by at least one order of magnitude; b) with the help of the new imaging devices and microcomputers, improved precision and the ability to achieve quantitative results can be made in real time; c) a smaller possibility that the interference pattern might be misinterpreted.

Parallel Advances in Fabrication

The advance in testing techniques would be incomplete without advances in fabrication techniques. Newer methods were developed, for example, fast polishing with high speed spindles and polyurethane laps. Computer controlled polishing began in the early '70s. Planetary machines for almost automatic polishing of high quality optical flats made it possible to manufacture them in much larger quantities. Teflon polishers have been developed to automatically polish optical flats with high quality, even better than we can measure them. Abrasive water cutting of glass permits us to cut glass like wood, with complicated shapes and quite fast. Ion bombardment polishing is now being used to aspherize surface with high reliability and accuracy. Another important fabrication development is the diamond turning technique.

All these new testing and fabrication techniques have made it possible to fabricate high quality optical components, with aspheric shapes that a few years ago were considered too complicated to even attempt to manufacture. However, it is important to point out that newer techniques are not a substitute for the high skill and training required by those in optics shops. Frequently, a high degree of ingenuity and imagination can produce better results than all automatic machines together.

OPTICAL DESIGN SOFTWARE

Optical design software allows the user to develop a configuration of optical elements that manipulate the trajectory of light for the purposes of creating an image, illuminating a target, coupling into a fiber, and so on. In many ways, it parallels the capabilities of popular CAD (computer-aided design) software for mechanical design but with a very significant difference: Optical design software can make changes to the optical elements to improve a performance metric. With enhanced usability possible in modern software interfaces, optical design software can make even an inexperienced lens designer capable of developing astonishingly well-performing optical systems.

Describing the Optical System

Interestingly enough, for historical reasons, most optical design software systems do not define the optical system as a collection of optical elements but rather as a collection of optical surfaces positioned in space. There is an almost dizzying variety of surface types: spheres, conics, aspheres, cones, cylinders, toroids, zernikes, NURBs, various types of polynomials, and so on. Virtually any geometrical surface can be represented. Additionally, the user can select from fictitious or idealized optical elements such as thin lenses, lenses represented by ABCD matrices or even simulated lenses based on eikonal theory that duplicate the aberration behavior of "real" lenses. However, even this expansive palette of surface types is insufficient for certain applications, and so in most cases the user can write customized code (usually in a higher-level language such as C or FORTRAN) to describe the surface and then link it to the optical design software. The imagination of the user is really the only limit when it comes to describing geometry.

The user then assigns to each of these surfaces optical properties such as dispersive homogeneous, gradient or birefringent refractive indices, and diffraction grating lines, as well as transmissive and reflective coatings. Once these are assigned, these surfaces become representations of lenses, mirrors, prisms, and other recognizable optical components and affect the trajectory of light passing through them according to the assigned properties. Figure shows the optical configuration of a double Gauss lens composed of two singlets and two doublets arranged around a central stop.

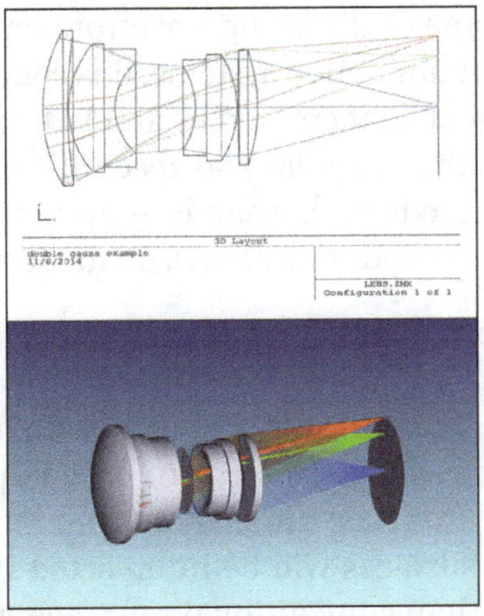

Cross section and 3D solid model of a double Gauss lens showing the trajectories of rays originating from three different field locations.

Although most optical systems have only a single detector and one path from the object to that detector, some systems employ beamsplitters and filters to create multiple optical paths addressing multiple detectors. Other systems such as classical zoom lenses have elements that move in space relative to each other to accomplish different optical tasks. These systems are called "multiconfiguration" or "zoom" systems, and users toggle between each of the configurations for both analysis and optimization as required.

Since custom optical elements can be expensive as well as not immediately available, most optical design software programs offer the user access to databases of commercial off-the-shelf (COTS) components available from numerous commercial vendors. Many successful prototype systems are developed by first designing with custom optical elements and then iteratively replacing the custom elements with readily available off-the-shelf parts, thereby taking advantage of the significant cost savings afforded by COTS.

Ray Tracing

The fundamental calculation for all optical design software is ray tracing: the calculation of the exact trajectory of a light ray through the optical system. Because the image-forming properties of an optical system can depend upon very small differences (fractions of a wavelength of light, in most cases) in the trajectories of rays, ray tracing is generally performed to the highest level of accuracy possible on the computer.

Rays are traced from the object to the image through all optical surfaces. Very often, the object emits light according to a spectrum, and so rays are traced at different wavelengths to characterize the chromatic properties of the system. Less commonly, the polarization of the ray is important to a given calculation; ray-trace algorithms have been extended to include the trajectory of polarized light through both homogenous and birefringent materials, as well as polarizing elements such as linear polarizers, quarter-wave and halfwave plates, and so on. In very specialized applications such as photolithography and stellar interferometry, distributions of rays representing partially spatially coherent light can also be traced through an optical system.

In classical optical design, light originates at a point located either at infinity or proximate to the optical system and therefore describes a plane or diverging wavefront entering the optical system. However, with the advent of optical design for illumination systems, this concept had to be extended; modern optical design software can now represent light emitted by an LED, arc source, thermal source or another incoherent source. Depending upon their angular and spectral emission characteristics, these types of sources can require upwards of hundreds to thousands — even millions — of rays. Surprisingly, even this number of rays can be traced in seconds or minutes on modern computers.

In traditional optical design, rays propagate from surface to surface in a sequential fashion, following the order in which the surfaces are defined. Again, to support optical design for illumination systems, it has been necessary to generalize the ray-trace algorithm to include "nonsequential" ray tracing, where the ray determines its trajectory not based upon the ordering of the surfaces in the software database, but instead from the dimensions and locations of the surfaces themselves. Since the rays do not know which surfaces will be intersected in which order, non-sequential ray tracing can be significantly slower than conventional ray tracing but represents the physical propagation of light through a generalized optical geometry.

Finally, there are applications where the use of geometrical rays fails to accurately describe the nature of the propagation of light through an optical system. In these cases, optical design software gives the user an alternative: a Fourier transform or beamlet-based propagation algorithm capable of modeling diffraction from apertures, interference, spatial filters, and other classical physical optics effects.

Optimization

Probably one of the most valuable features in modern optical design software is the ability to alter the prescription of the optical configuration to improve its performance for a given application. Broadly defined, this is the process of "optimization".

Optimization algorithms have been in development for over 50 years, and today's optical design software offers the user several powerful options. Most software offers a damped least-squares optimization algorithm that iteratively makes changes in the system to reduce the magnitude of a specified set of metrics; these metrics are collectively called the "error function," "defect function," "merit function" or similar nomenclature. However, the user may want to consider a larger parameter space (i.e., very dramatic departures from the starting point), so "global" optimization algorithms have been developed. These powerful tools can completely alter the initial optical system configuration; in some cases, they can add and remove elements in their search for a more optimum configuration. In most implementations, the algorithm searches for different solution spaces and then provides the user with a list of candidate designs from which the user can select and continue to refine.

As with optical surfaces, the variety of performance metrics that can be optimized is staggering: rms (root mean square) spot size, wavefront error, MTF (modulation transfer function), encircled energy, CIE color, irradiance uniformity, packaging, sensitivity to misalignment, beam clearance, ghost image suppression, tolerance to defocus, single-mode fiber coupling efficiency, local holographic grating frequency, cost, sensitivity to temperature changes, and so on. In those cases where the software lacks a specific metric to be optimized, the user can define one using construction and ray-trace database items.

Within the broad scope of optimization, most optical design software offers tools for specific applications. For example, most novice designers appreciate the ability of glass substitution/optimization tools to refine their glass selection; the refinement may be based upon image criteria but may also be related to cost, thermal expansion issues and availability, among others. Designers working with complex systems generally use software tools that interrogate the system looking for the optimum location(s) at which to place aspheric surface(s). Cam optimization software tells the zoom-lens designer exactly how groups of lens elements need to move to preserve image quality.

Analysis

There are numerous ways in which an optical system may be expected to perform, so it should not be surprising that there are numerous analysis options.

The simplest analysis option consists of diagnostics, usually of interest only to the optical designer. These include a single ray trace through the system, with the coordinates and direction of propagation (among other data) of the ray on each surface displayed. Aberration plots computed at specific locations across the field of view are very common in analysis, as these indicate the nature of the residual aberrations present in the system and their magnitudes. Distortion plots and astigmatic field curves often round out a typical analysis.

More specialized analysis options are also available. Polarization ray tracing can compute the orientation of the electric field vector as the light propagates through the optical system. Aberration coefficients frequently offer an insight into those surface responsible for high levels of aberration. Gaussian beams can be traced through the optical system to identify the locations of beam waists and astigmatic behavior. Multilayer coating properties such as reflection, transmission and absorption, among other properties, may be computed to compare with experimental data.

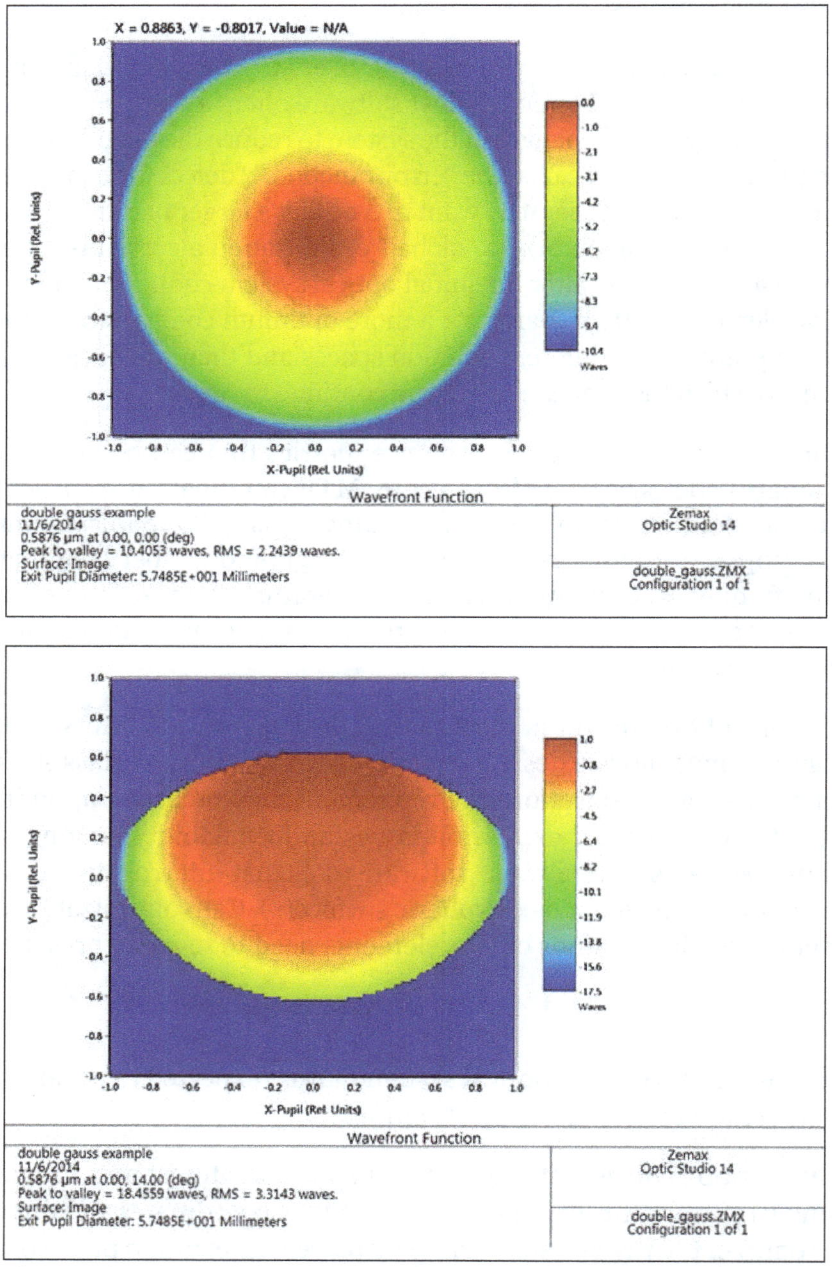

The wavefront at the exit pupil gives the lens designer a sense of the state of optical correction, as well as a visual clue of the amount of vignetting present as a function of field position.

Next-level refinement of analysis options consists of calculations and data manipulations involving hundreds or thousands of traced rays. Examples include spot diagrams (plots of ray intersection coordinates on the image plane), point-spread functions, encircled-energy calculations, and ghost-image analyses. An illumination engineer simulating an LED or arc source propagating through a hybrid optic might trace several million rays to ascertain the level of irradiance uniformity at some observation plane. Modern PCs are more than capable of tracing very large numbers of rays virtually instantaneously. Figure shows the exit-pupil wavefront for a double Gauss lens on-axis and the edge of the field.

Image analyses are probably the most sophisticated calculations commonly available in optical design software, and within this category are many different ways of evaluating imagery. For example, the designer may evaluate the MTF at several positions in the field across a spectral bandwidth. A designer working in the photolithographic industry might consider a partial spatial coherence resolution calculation to evaluate the resolution of a candidate system. In the photographic industry, a designer could perform an image simulation whereby an actual scene is imaged through an optical system to illustrate the resolution and residual distortion present in the final image.

Tolerancing and Fabrication Support

With the exception of theoretical design studies (such as the monochromatic quartet problem), the goal of most design efforts is to develop an optical prescription that can be fabricated at a reasonable cost. However, since fabrication is not an exact process, the resulting "as-built" radii, thicknesses, aspherics, refractive indices and so on are never exactly what the designer called for. Consequently, few fabricated systems deliver precisely the performance predicted by the optical design software. The challenge for the optical designer is to specify the degree to which each construction parameter must match its nominal value in the fabricated physical hardware while still delivering acceptable performance.

Since the exact departure from the design radius of curvature, for example, can never be known, tolerancing algorithms depend upon statistics to bound the problem. There are two common approaches found in virtually all optical design software. The first is a "brute force" statistical analysis where random perturbations are applied to specific individual construction parameters (i.e., a lens thickness), to groups of construction parameters (i.e., to all of the radii, thicknesses, refractive indices, etc., in a group of lenses) or to the entire system geometry. Each construction parameter may have its own distribution of perturbations. For example, opticians tend to make lenses somewhat thicker than called for in case the lens must be reworked. Radii of curvature, on the other hand, have equal probability of being convex or concave to the test plate. The optical design software may generate and apply hundreds to thousands of realizations of perturbations to the system, acquire the statistics and present to the designer a summary of the probability of meeting a given level of performance.

The second common approach to tolerancing is to follow this same process but, instead of generating actual perturbations, applying perturbations algebraically. These algorithms, usually built around rms wavefront and MTF metrics, are much faster and allow the designer to conduct tolerance studies in minutes, whereas a brute-force approach might take hours or days. Of course, all algorithms have limitations, so many designers employ the algebraic approach early in the tolerance process and then subsequently verify all tolerance assignments with a brute-force analysis.

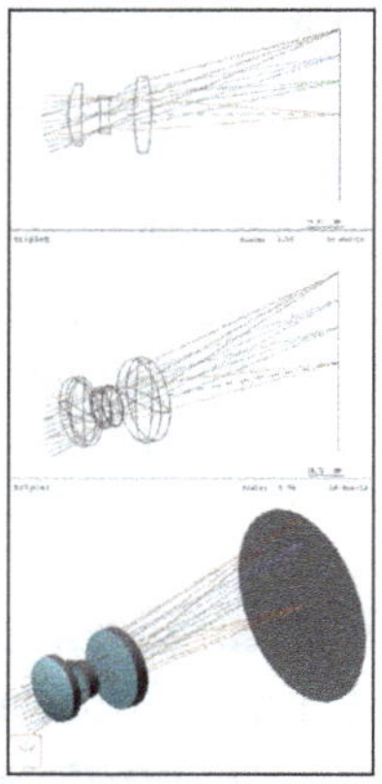

Example of increasing graphic sophistication, from a simple cross-sectional line plot to a 3D wireframe to a 3D solid model in modern lens design software.

The exact metric to be toleranced depends upon the application. A tolerance photographic objective may have to meet an MTF requirement. An astronomical telescope camera system working with CCD detector may have an encircled-energy requirement relative to the pixel dimensions. A reconnaissance camera could very well have a distortion tolerance. In an illumination system employing an LED and hybrid optic, the divergence and total power contained within the emitted beam may be the tolerance metric.

Ultimately, tolerancing is a trade-off between optical system performance and cost of fabrication. There is rarely a unique set of tolerances for a given application, and frequently a designer has no hard evidence that his/her tolerances are optimal. In many cases, the selection of tolerances is determined by experience, as well as by outputs from optical design software.

User Interface

Since traditional optical design involves the propagation of light sequentially through a collection of surfaces, most optical design software arranges the construction data in a spreadsheet fashion, where each row of data corresponds to a particular surface.

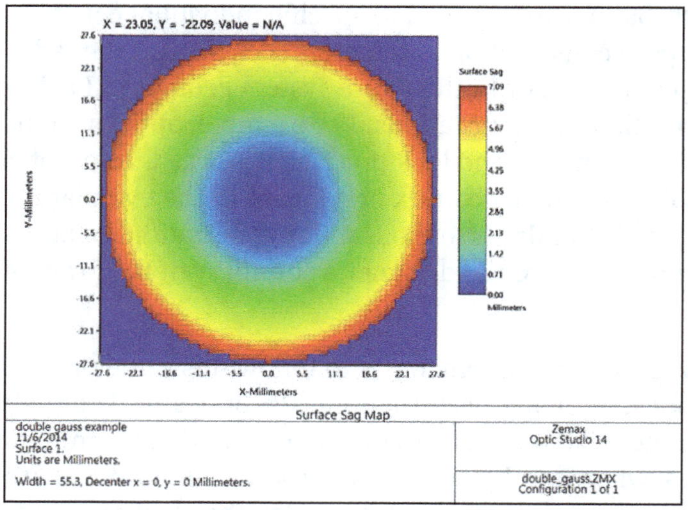

The sag map of a surface shows the departure from a plane. In the case of an aspheric surface, the magnitude of the departure could be the deciding issue in manufacturability.

Optical design software can provide the user with a seemingly endless number of graphics: two- and three-dimensional views of the optical system (including the ray traces, as shown in figure), transverse ray-aberration plots, wavefront plots, spot diagrams, point-spread function plots, MTF curves, encircled-energy plots, lateral chromatic-focus plots, Gaussian beam profile plots and so on. The optical designer likely must convey information to a broad audience (i.e., investor, client, fabricator, supplier, advertising agent, peer group, etc.), so software developers have made virtually all common analyses literally a push-button away. Figure shows the sag map of a surface; this information would be conveyed from the optical designer to the fabricator.

However, there are instances where what the optical designer does is simply a piece of a larger design/analysis effort. Accordingly, most optical design software has the ability to interface directly via DDE, .COM, or .NET to other, nonoptical applications such as MatLab, LabView, Mathematica, etc. In these cases, the optical design software frequently is queried for a specific calculation in response to an action taken by another piece of software involved in the analysis (e.g., a surface perturbation resulting from a thermal shock to the system). Simulations of this type can quickly become extraordinarily complex, but they provide the analyst with valuable information on how the system will perform in the real world.

References

- Frejlich, Jaime (2007). Photorefractive Materials: Fundamental Concepts, Holographic Recording and Materials Characterization. WILEY-INTERSCIENCE. ISBN 978-0-471-74866-3
- Optical+materials: thefreedictionary.com, Retrieved 5 June, 2019
- Blanche, P.A.; Tay, S. "An Updatable Holographic Three-Dimensional Display for 3D Visualization". Journal of Display Technology. 4: 424–430. Bibcode:2008jdist...4..424B. Doi:10.1109/jdt.2008.2001574
- Optical-glass, optics, application-notes: edmundoptics.eu, Retrieved 6 July 2019
- "Dr. Erhardt Glatzel (Biography)". The Zeiss Historica Society. Archived from the original on January 27, 2013. Retrieved July 21, 2013
- Optical-Design-Software-Fundamentals: photonics.com, Retrieved 7 August, 2019
- Fischer, Robert E.; Tadic-Galeb, Biljana; Yoder, Paul R. (2008). Optical System Design (2nd ed.). New York: mcgraw-Hill. Pp. 8, 179–198. ISBN 0-07-147248-7

Optical Instruments

CHAPTER 3

Some of the optical instruments are telescope, microscope, binoculars, camera, optical interferometer, photometer, polarimeter, refractometer, optical spectrometer, optical mapping, lasers and autocollimator. These diverse types of optical instruments have been thoroughly discussed in this chapter.

Optical instruments are based on scientific findings in field of optics. Optics is a branch of physics which examines and explains the travel of light and its interactions with matter. The principles of physical optics are included in the discipline of technical optics and therefore have an impact on construction, design and fabrication of optical instruments. The most famous optical instrument is of natural origin: the human eye. Its ability to convert electromagnetic waves with a wavelength of 380 nm (violet) to 780 nm (red), called visible light, by means of photosensitive cells in the retina, into neural impulses, which are evaluated in the visual cortex of the human brain, causes the humans ability of visual perception. The optic mechanism represented by the human eye is also used as the operating principle of simple optical instruments. By changing the curvature radius and refractive index, the focal distance can be varied and the beam of light can be focused, thus objects can be displayed with magnification. Microscopes, magnifying glasses, binoculars and telescopes are based on this simple principle. Mostly those instruments are passive devices, i.e. an external light source is necessary for the measuring. Complex systems are used in satellite techniques, where radio- and spectrometers are made for intensity measuring and spectral analysis. In addition to passive instruments, there are several areas where active optical instruments are used. For example, lasers are used in a CD-drive to read the stored data on the data carrier's surface. Another famous technology is the LIDAR (Light Detection and Ranging).

Active Optical Instruments.

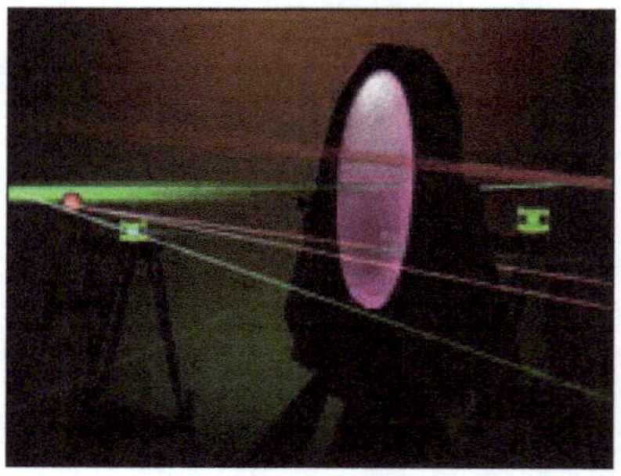

Passive Optical Instruments.

Functioning Principle of Optical Instruments

- Passive Optical Instruments: Important construction parts of passive optical instruments are optic lenses. Optic lenses are translucent glasses with two refractive surfaces. The refractive property of these surfaces is caused by concave or convex surface curvature, whereas a planar surface does not provoke any refraction of light at all. If an object is observed through a lens it can be either magnified or minimized depending on the implementation of the lens. Determining factors are mainly the focal distance and its reciprocal value as well as the index of refraction, with its unit dioptre, which is also important for classification of spectacle lenses. Therefore, the focal distances can be varied with the lenses in order to accommodate the optic image of an object in its size.

- Active optical instruments: In contrast to passive instruments, active optical instruments need their own light source to perform various tasks. The basic principle for this is the transmission of bundled electromagnetic waves and the reception of emerging echoes due to reflection on objects and surfaces. Whereas radar systems work with invisible short-wave radio waves, optical instruments send visible electromagnetic impulses (laser). Through relation of signal propagation time and light velocity, an accurate distance measurement can be performed, and the velocities can be determined in order to enable the imagery of objects.

TELESCOPE

Telescope is a device used to form magnified images of distant objects. The telescope is undoubtedly the most important investigative tool in astronomy. It provides a means of collecting and analyzing radiation from celestial objects, even those in the far reaches of the universe.

Galileo revolutionized astronomy when he applied the telescope to the study of extraterrestrial bodies in the early 17th century. Until then, magnification instruments had never been used for this purpose. Since Galileo's pioneering work, increasingly more powerful optical telescopes have been developed, as has a wide array of instruments capable of detecting and measuring radiation in every region of the electromagnetic spectrum. Observational capability has been further enhanced by the invention of various kinds of auxiliary instruments (e.g., the camera, spectrograph, and charge-coupled device) and by the use of electronic computers, rockets, and spacecraft in conjunction with telescope systems. These developments have contributed dramatically to advances in scientific knowledge about the solar system, the Milky Way Galaxy, and the universe as a whole.

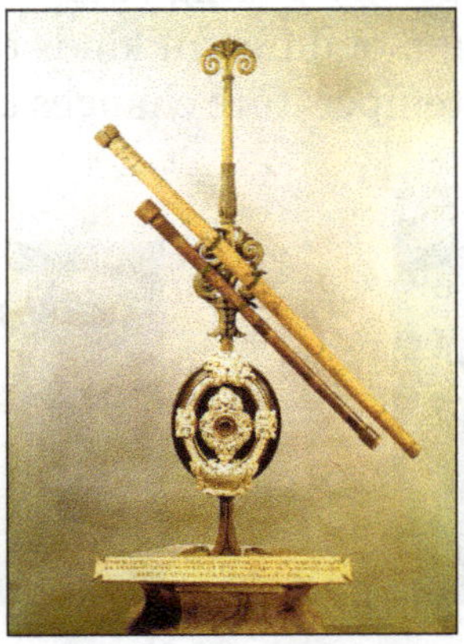

Two of Galileo's first telescopes; in the Institute and Museum of the History of Science, Florence.

Refracting Telescopes

Commonly known as refractors, telescopes of this kind are typically used to examine the Moon, other objects of the solar system such as Jupiter and Mars, and binary stars. The name refractor is derived from the term refraction, which is the bending of light when it passes from one medium to another of different density—e.g., from air to glass. The glass is referred to as a lens and may have one or more components. The physical shape of the components may be convex, concave, or plane-parallel. This diagram illustrates the principle of refraction and the term focal length. The focus is the point, or plane, at which light rays from infinity converge after passing through a lens and traveling a distance of one focal length. In a refractor the first lens through which light from a celestial object passes is called the objective lens. It should be noted that the light will be inverted at the focal plane. A second lens, referred to as the eyepiece lens, is placed behind the focal plane and enables the observer to view the enlarged, or magnified, image. Thus, the simplest form of refractor consists of an objective and an eyepiece.

The historical 91-cm (36-inch) refractor at the Lick Observatory.

The diameter of the objective is referred to as the aperture; it typically ranges from a few centimetres for small spotting telescopes up to one metre for the largest refractor in existence. The objective, as well as the eyepiece, may have several components. Small spotting telescopes may contain an extra lens behind the eyepiece to erect the image so that it does not appear upside-down. When an object is viewed with a refractor, the image may not appear sharply defined, or it may even have a predominant colour in it. Such distortions, or aberrations, are sometimes introduced when the lens is polished into its design shape. The major kind of distortion in a refractor is chromatic aberration, which is the failure of the differently coloured light rays to come to a common focus. Chromatic aberration can be minimized by adding components to the objective. In lens-design technology, the coefficients of expansion of different kinds of glass are carefully matched to minimize the aberrations that result from temperature changes of the telescope at night.

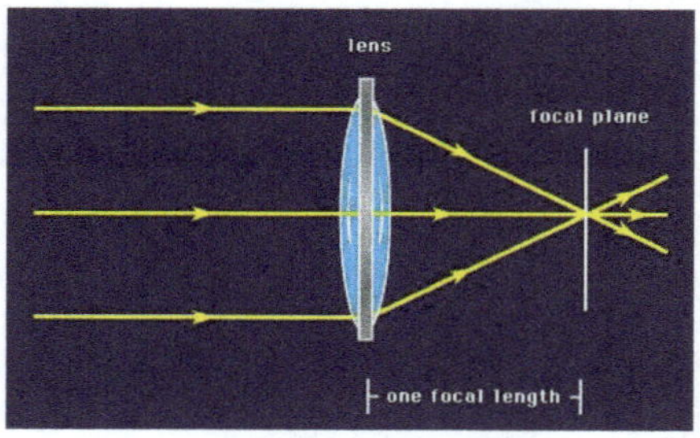

Focal length of a lens.

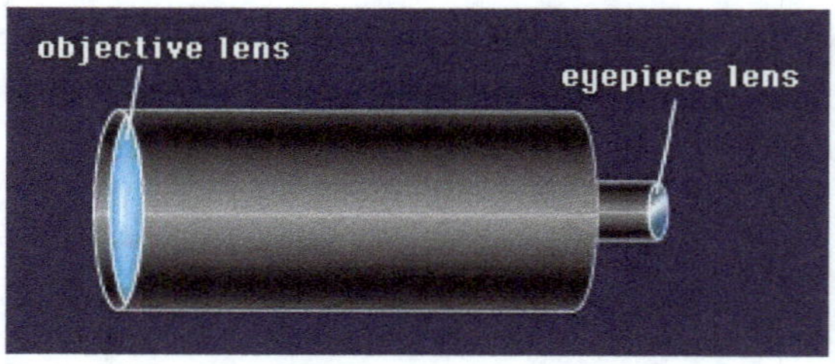

Refracting telescope.

Eyepieces, which are used with both refractors and reflector have a wide variety of applications and provide observers with the ability to select the magnification of their instruments. The magnification, sometimes referred to as magnifying power, is determined by dividing the focal length of the objective by the focal length of the eyepiece. For example, if the objective has a focal length of 254 cm (100 inches) and the eyepiece has a focal length of 2.54 cm (1 inch), then the magnification will be 100. Large magnifications are very useful for observing the Moon and the planets. However, since stars appear as point sources owing to their great distances, magnification provides no additional advantage when viewing them. Another important factor that one must take into consideration when attempting to view at high magnification is the stability of the telescope mounting. Any vibration in the mounting will also be magnified and may severely reduce the quality of the observed image. Thus, great care is usually taken to provide a stable platform for the telescope. This problem should not be associated with that of atmospheric seeing, which may introduce a disturbance to the image because of fluctuating air currents in the path of the light from a celestial or terrestrial object. Generally, most of the seeing disturbance arises in the first 30 metres (100 feet) of air above the telescope. Large telescopes are frequently installed on mountain peaks in order to get above the seeing disturbances.

Light Gathering and Resolution

The most important of all the powers of an optical telescope is its light-gathering power. This capacity is strictly a function of the diameter of the clear objective—that is, the aperture—of the telescope. Comparisons of different-sized apertures for their light-gathering power are calculated by the ratio of their diameters squared; for example, a 25-cm (10-inch) objective will collect four times the light of a 12.5-cm (5-inch) objective ($[25 \times 25] \div [12.5 \times 12.5] = 4$). The advantage of collecting more light with a larger-aperture telescope is that one can observe fainter stars, nebulae, and very distant galaxies.

Resolving power is another important feature of a telescope. This is the ability of the instrument to distinguish clearly between two points whose angular separation is less than the smallest angle that the observer's eye can resolve. The resolving power of a telescope can be calculated by the following formula: resolving power = 11.25 seconds of arc/d, where d is the diameter of the objective expressed in centimetres. Thus, a 25-cm-diameter objective has a theoretical resolution of 0.45 second of arc and a 250-cm (100-inch) telescope has one of 0.045 second of arc. An important application of resolving power is in the observation of visual binary stars. There, one star is routinely observed as it revolves around a second star. Many observatories conduct extensive visual binary observing programs and publish catalogs of their observational results. One of the major contributors in this field is the United States Naval Observatory in Washington, D.C.

Most refractors currently in use at observatories have equatorial mountings. The mounting describes the orientation of the physical bearings and structure that permits a telescope to be pointed at a celestial object for viewing. In the equatorial mounting, the polar axis of the telescope is constructed parallel to Earth's axis. The polar axis supports the declination axis of the instrument. Declination is measured on the celestial sky north or south from the celestial equator. The declination axis makes it possible for the telescope to be pointed at various declination angles as the instrument is rotated about the polar axis with respect to right ascension. Right ascension is measured along the celestial equator from the vernal equinox (i.e., the position on the celestial sphere where the Sun crosses the celestial equator from south to north on the first day of spring). Declination and right ascension are the two coordinates that define a celestial object on the celestial sphere. Declination is analogous to latitude, and right ascension is analogous to longitude. Graduated dials are mounted on the axis to permit the observer to point the telescope precisely. To track an object, the telescope's polar axis is driven smoothly by an electric motor at a sidereal rate—namely, at a rate equal to the rate of rotation of Earth with respect to the stars. Thus, one can track or observe with a telescope for long periods of time if the sidereal rate of the motor is very accurate. High-accuracy motor-driven systems have become readily available with the rapid advancement of quartz-clock technology. Most major observatories now rely on either quartz or atomic clocks to provide accurate sidereal time for observations as well as to drive telescopes at an extremely uniform rate.

A notable example of a refracting telescope is the 66-cm (26-inch) refractor of the U.S. Naval Observatory. This instrument was used by the astronomer Asaph Hall to discover the two moons of Mars, Phobos and Deimos, in 1877. Today, the telescope is used primarily for observing binary stars. The 91-cm (36-inch) refractor at Lick Observatory on Mount Hamilton, Calif., U.S., and the 1-metre (40-inch) instrument at Yerkes Observatory in Williams Bay, Wis., U.S., are the largest refracting systems currently in operation.

Another type of refracting telescope is the astrograph, which usually has an objective diameter of approximately 20 cm (8 inches). The astrograph has a photographic plateholder mounted in the focal plane of the objective so that photographs of the celestial sphere can be taken. The photographs are usually taken on glass plates. The principal application of the astrograph is to determine the positions of a large number of faint stars. These positions are then published in catalogs such as the *AGK3* and serve as reference points for deep-space photography.

Reflecting Telescopes

Reflectors are used not only to examine the visible region of the electromagnetic spectrum but also to explore both the shorter- and longer-wavelength regions adjacent to it (i.e., the ultraviolet and the infrared). The name of this type of instrument is derived from the fact that the primary mirror reflects the light back to a focus instead of refracting it. The primary mirror usually has a concave spherical or parabolic shape, and, as it reflects the light, it inverts the image at the focal plane. The formulas for resolving power, magnifying power, and light-gathering power, as discussed for refractors, apply to reflectors as well.

The 72-inch reflecting telescope at Birr Castle was the largest in the world at the time of its construction.

The primary mirror is located at the lower end of the telescope tube in a reflector and has its front surface coated with an extremely thin film of metal, such as aluminum. The back of the mirror is usually made of glass, although other materials have been used from time to time. Pyrex was the principal glass of choice for many of the older large telescopes, but new technology has led to the development and widespread use of a number of glasses with very low coefficients of expansion. A low coefficient of expansion means that the shape of the mirror will not change significantly as the temperature of the telescope changes during the night. Since the back of the mirror serves only to provide the desired form and physical support, it does not have to meet the high optical quality standards required for a lens.

Reflecting telescopes have a number of other advantages over refractors. They are not subject to chromatic aberration because reflected light does not disperse according to wavelength. Also, the telescope tube of a reflector is shorter than that of a refractor of the same diameter, which reduces the cost of the tube. Consequently, the dome for housing a reflector is smaller and more economical to construct. So far only the primary mirror for the reflector has been discussed. One might wonder about the location of the eyepiece. The primary mirror reflects the light of the celestial object to the prime focus near the upper end of the tube. Obviously, if an observer put his eye there to observe with a modest-sized reflector, he would block out the light from the primary mirror with his head. Isaac Newton placed a small plane mirror at an angle of 45° inside the prime focus and thereby brought the focus to the side of the telescope tube. The amount of light lost by this procedure is very small when compared to the total light-gathering power of the primary mirror. The Newtonian reflector is popular among amateur telescope makers.

Isaac Newton's reflecting telescope.

Laurent Cassegrain of France, a contemporary of Newton, invented another type of reflector. Called the Cassegrainian telescope, this instrument employs a small convex mirror to reflect the light back through a small hole in the primary mirror to a focus located behind the primary. The diagram illustrates a typical Cassegrain reflector. Some large telescopes of this kind do not have a hole in the primary mirror but use a small plane mirror in front of the primary to reflect the light outside the main tube and provide another place for observation. The Cassegrain design usually permits short tubes relative to their mirror diameter.

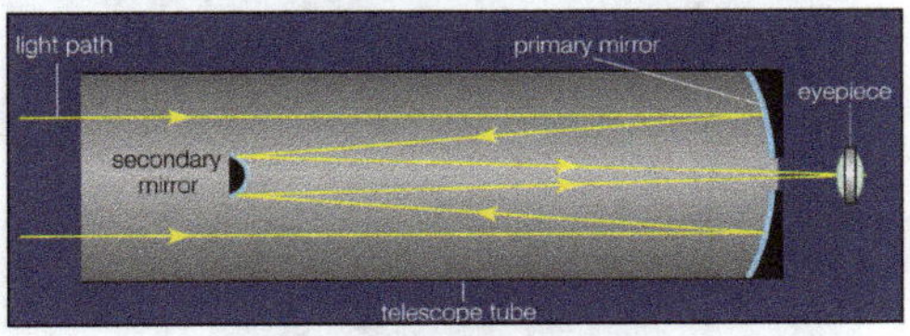

Cassegrain reflector.

One more variety of reflector was invented by another of Newton's contemporaries, the Scottish astronomer James Gregory. Gregory placed a concave secondary mirror outside the prime focus to reflect the light back through a hole in the primary mirror. Notable is the fact that the Gregorian design was adopted for the Earth-orbiting space observatory, the Solar Maximum Mission (SMM), launched in 1980.

Most large reflecting telescopes that are currently in use have a cage at their prime focus that permits the observer to sit inside the telescope tube while operating the instrument. The 5-metre (200-inch) reflector at Palomar Observatory, near San Diego, Calif., is so equipped. While most reflectors have equatorial mounts similar to refractors, the world's largest reflector, the 10.4-metre (34.1-foot) instrument at the Gran Telescopio Canarias at La Palma, Canary Islands, Spain, has an altitude-azimuth mounting. The significance of the latter design is that the telescope must be moved both in altitude and in azimuth as it tracks a celestial object. Equatorial mountings, by contrast, require motion in only one coordinate while tracking, since the declination coordinate is constant. Reflectors, like refractors, usually have small guide telescopes mounted parallel to their main optical axis to facilitate locating the desired object. These guide telescopes have low magnification and a wide field of view, the latter being a desirable attribute for finding stars or other remote cosmic objects.

The parabolic shape of a primary mirror has a basic failing in that it produces a narrow field of view. This can be a problem when one wishes to observe extended celestial objects. To overcome this difficulty, most large reflectors now have a modified Cassegrain design. The central area of the primary mirror has its shape deepened from that of a paraboloid, and the secondary mirror is configured to compensate for the altered primary. The result is the Ritchey-Chrétien design, which has a curved rather than a flat focus. Obviously, the photographic medium must be curved to collect high-quality images across the curved focal plane. The 1-metre telescope of the U.S. Naval Observatory in Flagstaff, Ariz., was one of the early examples of this design.

The Schmidt Telescope

The Ritchey-Chrétien design has a good field of view of about 1°. For some astronomical applications, however, photographing larger areas of the sky is mandatory. In 1930 Bernhard Schmidt, an optician at the Hamburg Observatory in Bergedorf, Ger., designed a catadioptric telescope that satisfied the requirement of photographing larger celestial areas. A catadioptrictelescope design incorporates the best features of both the refractor and the reflector—i.e., it has both reflective and refractive optics. The Schmidt telescope has a spherically shaped primary mirror. Since parallel light rays that are reflected by the centre of a spherical mirror are focused farther away than those reflected from the outer regions, Schmidt introduced a thin lens (called the correcting plate) at the radius of curvature of the primary mirror. Since this correcting plate is very thin, it introduces little chromatic aberration. The resulting focal plane has a field of view several degrees in diameter. The diagram illustrates a typical Schmidt design.

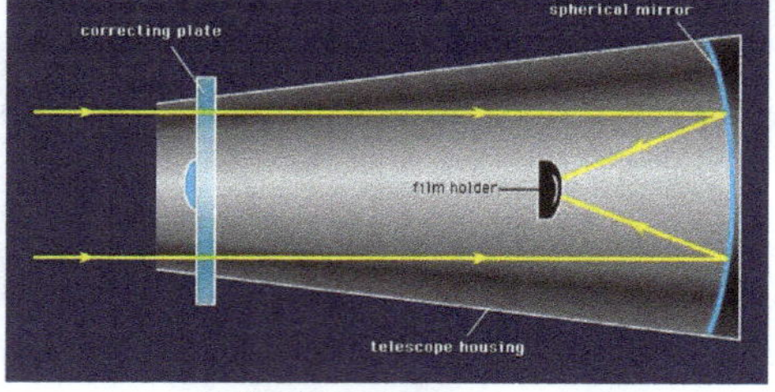

Schmidt telescope.

The National Geographic Society–Palomar Observatory Sky Survey made use of a 1.2-metre (47-inch) Schmidt telescope to photograph the northern sky in the red and blue regions of the visible spectrum. The survey produced 900 pairs of photographic plates (about 7° by 7° each) taken from 1949 to 1956. Schmidt telescopes of the European Southern Observatory in Chile and of the Siding Spring Observatory in Australia have photographed the remaining part of the sky that cannot be observed from Palomar Observatory. (The survey undertaken at the latter included photographs in the infrared as well as in the red and blue spectral regions).

Multimirror Telescopes

The main reason astronomers build larger telescopes is to increase light-gathering power so that they can see deeper into the universe. Unfortunately, the cost of constructing larger single-mirror telescopes increases rapidly—approximately with the cube of the diameter of the aperture. Thus, in order to achieve the goal of increasing light-gathering power while keeping costs down, it has become necessary to explore new, more economical and nontraditional telescope designs.

The two 10-metre (33-foot) Keck Observatory multimirror telescopes represent such an effort. The first was installed on Mauna Kea on the island of Hawaii in 1992, and a second telescope was completed in 1996. Each of the Keck telescopes comprises 36 contiguous adjustable mirror segments, all under computer control. Even-larger multimirror instruments are currently being planned by American and European astronomers.

Mauna Kea Observatory: Keck telescopes.

Special Types of Optical Telescopes

Solar Telescopes

Either a refractor or a reflector may be used for visual observations of solar features, such as sunspots or solar prominences. Special solar telescopes have been constructed, however, for investigations of the Sun that require the use of such ancillary instruments as spectroheliographs and coronagraphs. These telescopes are mounted in towers and have very long focus objectives. Typical examples of tower solar telescopes are found at the Mount Wilson Observatory in California and the McMath-Hulbert Observatory in Michigan. The long focus objective produces a very good scale factor, which in turn makes it possible to look at individual wavelengths of the solar electromagnetic spectrum in great detail. A tower telescope has an equatorially mounted plane mirror at its summit to direct the sunlight into the telescope objective. This plane mirror is called a coelostat. Bernard Lyot constructed another type of solar telescope in 1930 at Pic du Midi Observatory in

France. This instrument was specifically designed for photographing the Sun's corona (the outer layer), which up to that time had been successfully photographed only during solar eclipses. The coronagraph, as this special telescope is called, must be located at a high altitude to be effective. The high altitude is required to reduce the scattered sunlight, which would reduce the quality of the photograph. The High Altitude Observatory in Colorado and the Sacramento Peak Observatory in New Mexico have coronagraphs. Coronagraphs are also used on board satellites, such as the Solar and Heliospheric Observatory, that study the Sun.

The Sun's corona as seen by the Large Angle Spectrometric Coronagraph (LASCO) aboard the Solar and Heliospheric Observatory (SOHO).

Earth-orbiting Space Telescopes

Cutaway of the Hubble Space Telescope, revealing the Optical Telescope Assembly (OTA), the heart of this orbiting observational system.

While astronomers continue to seek new technological breakthroughs with which to build larger ground-based telescopes, it is readily apparent that the only solution to some scientific problems is to make observations from above Earth's atmosphere. A series of Orbiting Astronomical Observatories (OAOs) was launched by the National Aeronautics and Space Administration (NASA). The OAO launched in 1972—later named Copernicus—had an 81-cm (32-inch) telescope on board. The most sophisticated observational system placed in Earth orbit so far is the Hubble Space Telescope. Launched in 1990, the HST is essentially a telescope with a 2.4-metre (94-inch) primary mirror. It has been designed to enable astronomers to see into a volume of space 300 to 400 times larger than that permitted by other systems. At the same time, the HST is not impeded by any of the problems caused by the atmosphere. It is equipped with five principal scientific instruments: (1) a wide-field and planetary camera, (2) a faint-object spectrograph, (3) a high-resolution spectrograph, (4) a high-speed photometer, and (5) a faint-object camera. The HST was launched into orbit from the U.S. space shuttle at an altitude of more than 570 km (350 miles) above Earth. Shortly after its deployment in Earth orbit, HST project scientists found that a manufacturing error affecting the shape of the telescope's primary mirror severely impaired the instrument's focusing capability. The flawed mirror caused spherical aberration, which limited the ability of the HST to distinguish between cosmic objects that lie close together and to image distant galaxies and quasars. Project scientists devised measures that enabled them to compensate for the defective mirror and correct the imaging problem.

Astronomical Transit Instruments

These small but extremely important telescopes have played a vital role in mapping the celestial sphere. Astronomical transit instruments are usually refractors with apertures of 15 to 20 cm (6 to 8 inches). (Ole Rømer, a Danish astronomer, is credited with having invented this type of telescope system.) The main optical axis of the instrument is aligned on a north-south line such that its motion is restricted to the plane of the meridian of the observer. The observer's meridian is a great circle on the celestial sphere that passes through the north and south points of the horizon as well as through the zenith of the observer. Restricting the telescope to motion only in the meridian provides an added degree of stability, but it requires the observer to wait for the celestial object to rotate across his meridian. The latter process is referred to as transiting the meridian, from which the name of the telescope is derived. There are various types of transit instruments—for example, the transit circle telescope, the vertical circle telescope, and the horizontal meridian circle telescope. The transit circle determines the right ascension of celestial objects, while the vertical circle measures only their declinations. Transit circles and horizontal meridian circles measure both right ascension and declination at the same time. The final output data of all transit instruments are included in star or planetary catalogs.

One of the most accurate astronomical transit instruments in the world is the U.S. Naval Observatory's 15-cm transit circle telescope. Other notable examples of this class of telescopes include the transit circle of the National Astronomical Observatory in Tokyo, the meridian circle of the Bordeaux Observatory in France, and the automatic meridian circle of the Roque de los Muchachos Observatory in the Canary Islands.

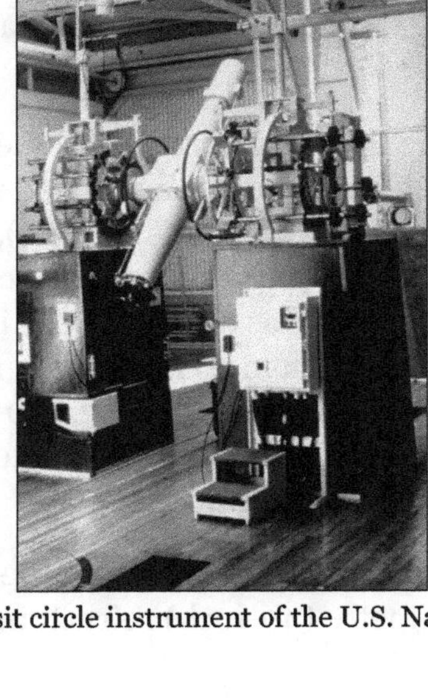

The 15-cm transit circle instrument of the U.S. Naval Observatory.

Astrolabes

Another special type of telescopic instrument is the modern version of the astrolabe. Known as a prismatic astrolabe, it too is used for making precise determinations of the positions of starsand planets. It may sometimes be used inversely to determine the latitude and longitude of the observer, assuming the star positions are accurately known. The aperture of a prismatic astrolabe is small, usually only 8 to 10 cm (3 to 4 inches). A small pool of mercury and a refracting prism make up the other principal parts of the instrument. An image reflected off the mercury is observed along with a direct image to give the necessary position data. The most notable example of this type of instrument is the French-constructed Danjon astrolabe. During the 1970s, however, the Chinese introduced various innovations that resulted in a more accurate and automatic kind of astrolabe, which is now in use at the National Astronomical Observatories of China's headquarters in Beijing.

The Development of the Telescope and Auxiliary Instrumentation

Evolution of the Optical Telescope

Galileo is credited with having developed telescopes for astronomical observation in 1609. While the largest of his instruments was only about 120 cm (47 inches) long and had an objective diameter of 5 cm (2 inches), it was equipped with an eyepiece that provided an upright (i.e., erect) image. Galileo used his modest instrument to explore such celestial phenomena as the valleys and mountains of the Moon, the phases of Venus, and the four largest Jovian satellites, which had never been systematically observed before.

The reflecting telescope was developed in 1668 by Newton, though John Gregory had independently conceived of an alternative reflector design in 1663. Cassegrain introduced another variation of the reflector in 1672. Near the end of the century, others attempted to construct refractors as long as 61 metres, but these instruments were too awkward to be effective.

The most significant contribution to the development of the telescope in the 18th century was that of Sir William Herschel. Herschel, whose interest in telescopes was kindled by a modest 5-cm Gregorian, persuaded the king of England to finance the construction of a reflector with a 12-metre (39-foot) focal length and a 120-cm mirror. Herschel is credited with having used this instrument to lay the observational groundwork for the concept of extragalactic "nebulas"—i.e., galaxies outside the Milky Way system.

Reflectors continued to evolve during the 19th century with the work of William Parsons, 3rd earl of Rosse, and William Lassell. In 1845 Lord Rosse constructed in Ireland a reflector with a 185-cm (73-inch) mirror and a focal length of about 16 metres (52 feet). For 75 years this telescope ranked as the largest in the world and was used to explore thousands of nebulae and star clusters. Lassell built several reflectors, the largest of which was on Malta; this instrument had a 124-cm (49-inch) primary mirror and a focal length of more than 10 metres (33 feet). His telescope had greater reflecting power than Rosse's, and it enabled him to catalog 600 new nebulae as well as to discover several satellites of the outer planets—Triton (Neptune's largest moon), Hyperion (Saturn's 8th moon), and Ariel and Umbriel (two of Uranus's moons).

Refractor telescopes, too, underwent development during the 18th and 19th centuries. The last significant one to be built was the 1-metre (40-inch) refractor at Yerkes Observatory. Installed in

1897, it remains the largest refracting system in the world. Its objective was designed and constructed by the optician Alvan Clark, while the mount was built by the firm of Warner & Swasey.

The reflecting telescope predominated in the 20th century. The rapid proliferation of increasingly larger instruments of this type began with the installation of the 2.5-metre (60-inch) reflector at the Mount Wilson Observatory near Pasadena, Calif., U.S. The technology for mirrors underwent a major advance when the Corning Glass Works (in Steuben county, N.Y., U.S.) developed Pyrex. This borosilicate glass, which undergoes substantially less expansion than ordinary glass does, was used in the 5-metre (200-inch) Hale Telescope built in 1948 at the Palomar Observatory. Pyrex also was utilized in the main mirror of the 6-metre (236-inch) reflector of the Special Astrophysical Observatory in Zelenchukskaya, Russia. Since then, much better materials for mirrors have become available. Cer-Vit, for example, was used for the 4.2-metre (165-inch) William Herschel Telescope of the Roque de los Muchachos Observatory in the Canary Islands, and Zerodur was used for the 3.5-metre (138-inch) reflector at the German-Spanish Astronomical Center in Calar Alto, Spain.

Advances in Auxiliary Instrumentation

Almost as important as the telescope itself are the auxiliary instruments that the astronomer uses to exploit the light received at the focal plane. Examples of such instruments are the camera, spectrograph, photomultiplier tube, charge-coupled device (CCD), and charge injection device (CID). Each of these instrument types is discussed below.

Cameras

American John Draper photographed the Moon as early as 1840 by applying the daguerreotype process. The French physicists A.-H.-L. Fizeau and J.-B.-L. Foucault succeeded in making a photographic image of the Sun in 1845. Five years later astronomers at Harvard Observatory took the first photographs of the stars.

The use of photographic equipment in conjunction with telescopes has benefited astronomers greatly, giving them two distinct advantages: first, photographic images provide a permanent record of celestial phenomena, and, second, photographic plates integrate the light from celestial sources over long periods of time and thereby permit astronomers to see much-fainter objects than they would be able to observe visually. Typically, the camera's photographic plate (or film) is mounted in the focal plane of the telescope. The plate or film consists of glass or of a plastic material that is covered with a thin layer of a silver compound. The light striking the photographic medium causes the silver compound to undergo a chemical change. When processed, a negative image results—i.e., the brightest spots (the Moon and the stars, for example) appear as the darkest areas on the plate or the film. Since the 1980s the CCD has supplanted photography in the production of astronomical images.

Spectrographs

Newton noted the interesting way in which a piece of glass can break up light into different bands of colour, but it was not until 1814 that the German physicist Joseph von Fraunhoferdiscovered the lines of the solar spectrum and laid the basis for spectroscopy. The spectrograph consists of a slit, a collimator, a prism for dispersing the light, and a focusing lens. The collimator is an optical device that produces parallel rays from a focal plane source—i.e., it gives the appearance that the source is located at an infinite distance. The spectrograph enables astronomers to analyze the chemical composition of planetary atmospheres, stars, nebulae, and other celestial objects. A bright line in the spectrum indicates the presence of a glowing gas radiating at a wavelength characteristic of the chemical element in the gas. A dark line in the spectrum usually means that a cooler gas has intervened and absorbed the lines of the element characteristic of the intervening material. The lines may be displaced to either the red end or the blue end of the spectrum. This effect was first noted in 1842 by the Austrian physicist Christian Johann Doppler. When a light source is approaching, the lines are shifted toward the blue end of the spectrum, and when the source is receding, the lines are shifted toward its red end. This effect, known as the Doppler effect, permits astronomers to study the relative motions of celestial objects with respect to Earth's motion.

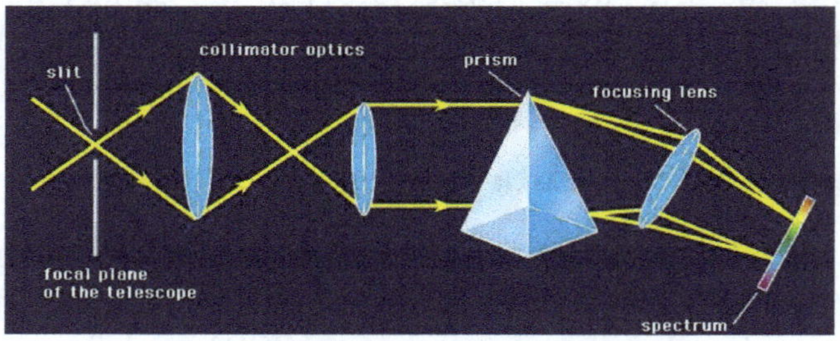

A spectrograph and its components.

The slit of the spectrograph is placed at the focal plane of the telescope. The resulting spectrum may be recorded photographically or with some kind of electronic detector, such as a photomultiplier tube, CCD, or CID. If no recording device is used, then the optical device is technically referred to as a spectroscope.

Photomultiplier Tubes

The photomultiplier tube is an enhanced version of the photocell, which was first used by astronomers to record data electronically. The photocell contains a photosensitive surface that generates an electric current when struck by light from a celestial source. The photosensitive surface is positioned just behind the focus. A diaphragm of very small aperture is usually placed in the focal plane to eliminate as much of the background light of the sky as possible. A small lens is used to focus the focal plane image on the photosensitive surface, which, in the case of a photomultiplier tube, is referred to as the photocathode. In the photomultiplier tube a series of special sensitive plates are arranged geometrically to amplify or multiply the electron stream. Frequently, magnifications of a million are achieved by this process.

The photomultiplier tube has a distinct advantage over the photographic plate. With the photographic plate the relationship between the brightness of the celestial source and its registration on the plate is not linear. In the case of the photomultiplier tube, however, the release of electrons in the tube is directly proportional to the intensity of light from the celestial source. This linear relationship is very useful for working over a wide range of brightness. A disadvantage of the photomultiplier tube is that only one object can be recorded at a time. The output from such a device is sent to a recorder or digital storage device to produce a permanent record.

Charge-coupled Devices

The charge-coupled device (CCD) uses a light-sensitive material on a silicon chip to electronically detect photons in a way similar to the photomultiplier tube. The principal difference is that the chip also contains integrated microcircuitry required to transfer the detected signal along a row of discrete picture elements (or pixels) and thereby scan a celestial object or objects very rapidly. When individual pixels are arranged simply in a single row, the detector is referred to as a linear array. When the pixels are arranged in rows and columns, the assemblage is called a two-dimensional array.

Pixels can be assembled in various sizes and shapes. The Hubble Space Telescope has a CCD detector with a 1,600 × 1,600 pixel array. Actually, there are four 800 × 800 pixel arrays mosaicked together. The sensitivity of a CCD is 100 times greater than a photographic plate and so has the ability to quickly scan objects such as planets, nebulae, and star clusters and record the desired data. Another feature of the CCD is that the detector material may be altered to provide more sensitivity at different wavelengths. Thus, some detectors are more sensitive in the blue region of the spectrum than in the red region.

Today, most large observatories use CCDs to record data electronically. Another similar device, the charge injection device, is sometimes employed. The basic difference between the CID and the CCD is in the way the electric charge is transferred before it is recorded; however, the two devices may be used interchangeably as far as astronomical work is concerned.

Impact of Technological Developments

Computers

Besides the telescope itself, the electronic computer has become the astronomer's most important tool. Indeed, the computer has revolutionized the use of the telescope to the point where the collection of observational data is now completely automated. The astronomer need only identify the object to be observed, and the rest is carried out by the computer and auxiliary electronic equipment.

A telescope can be set to observe automatically by means of electronic sensors appropriately placed on the telescope axis. Precise quartz or atomic clocks send signals to the computer, which in turn activates the telescope sensors to collect data at the proper time. The computer not only makes possible more efficient use of telescope time but also permits a more detailed analysis of the data collected than could have been done manually. Data analysis that would have taken a lifetime or longer to complete with a mechanical calculator can now be done within hours or even minutes with a high-speed computer.

Improved means of recording and storing computer data also have contributed to astronomical research. Optical disc data-storage technology, such as the CD-ROM (compact disc read-only memory) or the DVD-ROM (digital video disc read-only memory), has provided astronomers with the ability to store and retrieve vast amounts of telescopic and other astronomical data.

Rockets and Spacecraft

The quest for new knowledge about the universe has led astronomers to study electromagnetic radiation other than just visible light. Such forms of radiation, however, are blocked for the most part by Earth's atmosphere, and so their detection and analysis can only be achieved from above this gaseous envelope.

During the late 1940s, single-stage sounding rockets were sent up to 160 km (100 miles) or more to explore the upper layers of the atmosphere. From 1957, more sophisticated multistage rockets were launched as part of the International Geophysical Year. These rockets carried artificial satellites equipped with a variety of scientific instruments. Beginning in 1959 the Soviet Union and the United States, engaged in a "space race," intensified their efforts and launched a series of unmanned probes to explore the Moon. Lunar exploration culminated with the first manned landing on the Moon, by the U.S. Apollo 11 astronauts on July 20, 1969. Numerous other U.S. and Soviet spacecraft were sent to further study the lunar environmentuntil the mid-1970s. Lunar exploration revived in the early years of the 21st century with the United States, China, Japan, and India all sending unmanned probes to the Moon.

Starting in the early 1960s both the United States and the Soviet Union launched a multitude of unmanned deep-space probes to learn more about the other planets and satellites of the solar system. Carrying television cameras, detectors, and an assortment of other instruments, these probes sent back impressive amounts of scientific data and close-up pictures. Among the most successful missions were those involving the U.S. Messenger flybys of Mercury, the Soviet Venera probes to Venus the U.S. Mars Exploration Rover landings on Mars, and the U.S. Voyager 2 flybys of Jupiter, Saturn, Uranus, and Neptune. When the Voyager 2 probe flew past Neptune and its moons in August 1989, every known major planet had been explored by spacecraft. Many long-held views, particularly those about the outer planets, were altered by the findings of the Voyager probe. These findings included the discovery of several rings and six additional satellites around Neptune, all of which are undetectable to ground-based telescopes.

Artist's conception of Mars Exploration Rover.

Specially instrumented spacecraft have enabled astronomers to investigate other celestial phenomena as well. The Orbiting Solar Observatories and Solar Maximum Mission (Earth-orbiting U.S. satellites equipped with ultraviolet detector systems) have provided a means for studying solar activity. Another example is the Giotto probe of the European Space Agency, which enabled astronomers to obtain detailed photographs of the nucleus of Halley's Comet-during its 1986 passage.

MICROSCOPE

A microscope is an instrument used to produce enlarged images of small objects. The most common kind of microscope is an optical microscope, which uses lenses to form images from visible light. Electron microscopes form images from beams of electrons. Acoustic microscopes form images from high-frequency sound waves. Tunneling microscopes form images from the ability of electrons to "tunnel" through the surface of solids at extremely small distances.

An optical microscope with a single lens is known as a simple microscope. Simple microscopes include magnifying glasses and jeweler's loupes. An optical microscope with two lenses is known as a compound microscope. The basic parts of a compound microscope are the objective, which holds the lens near the specimen, and the eyepiece, which holds the lens near the observer. A modern compound microscope also includes a source of light (either a mirror to catch external light or a light bulb to provide internal light), a focusing mechanism, and a stage (a surface on which the object being examined can be held in place). Compound microscopes may also include a built-in camera for microphotography.

Ancient peoples noted that objects seen through water appeared larger. The first century Roman philosopher Seneca recorded the fact that letters seen through a glass globe full of water were magnified. The earliest simple microscopes consisted of a drop of water captured in a small hole in a piece of wood or metal. During the Renaissance, small glass lenses replaced the water. By the late seventeenth century, the Dutch scientist Antonie van Leeuwenhoek built outstanding simple microscopes using very small, high-quality lenses mounted between thin brass plates. Because of the excellence of his microscopes, and the fact that he was the first to make observations of microscopic organisms, Leeuwenhoek is often incorrectly thought of as the inventor of the microscope.

The compound microscope made its first appearance between the years 1590 and 1608. Credit for this invention is often given to Hans Janssen, his son Zacharias Janssen, or Hans Lippershey, all of whom were Dutch spectacle makers. Early compound microscopes consisted of pairs of lenses held in a small metal tube and looked much like modern kaleidoscopes. Because of the problem of chromatic aberration (the tendency of a lens to focus each color of light at a slightly different point, leading to a blurred image) these microscopes were inferior to well-made simple microscopes of the time.

The earliest written records of microscopic observations were made by the Italian scientist Francesco Stelluti in 1625, when he published drawings of a bee as seen through a microscope. The first drawings of bacteria were made by Leeuwenhoek in 1683. During the seventeenth and eighteenth centuries, numerous mechanical improvements were made in microscopes in Italy, including focusing devices and devices for holding specimens in place. In England in 1733, the amateur optician Chester Moor Hall discovered that combining two properly shaped lenses made of two different kinds of glass minimized chromatic aberration. In 1774, Benjamin Martin used this technique in a microscope. Many advances were made in the building of microscopes in the nineteenth and twentieth centuries. Electron microscopes were developed in the 1930s, acoustic microscopes in the 1970s, and tunneling microscopes in the 1980s.

Raw Materials

An optical microscope consists of an optical system (the eyepiece, the objective, and the lenses inside them) and hardware components which hold the optical system in place and allow it to be adjusted and focused. An inexpensive microscope may have a mirror as a light source, but most professional microscopes have a built-in light bulb.

Lenses are made of optical glass, a special kind of glass which is much purer and more uniform than ordinary glass. The most important raw material in optical glass is silicon dioxide, which must be more than 99.9% pure. The exact optical properties of the glass are determined by its other ingredients.

These may include boron oxide, sodium oxide, potassium oxide, barium oxide, zinc oxide, and lead oxide. Lenses are given an antireflective coating, usually of magnesium fluoride.

The eyepiece, the objective, and most of the hardware components are made of steel or steel and zinc alloys. A child's microscope may have an external body shell made of plastic, but most microscopes have an body shell made of steel.

If there is a mirror included, it is usually made of a strong glass such as Pyrex (a trade name for a glass made from silicon dioxide, boron dioxide, and aluminum oxide). The mirror has a reflective coating made of aluminum and a protective coating made of silicon dioxide.

If a light bulb is included, it is made from glass and contains a tungsten filament and wires made of nickel and iron within a mixture of argon and nitrogen gases. The base of the light bulb is made of aluminum.

If a camera is included, it contains lenses made of optical glass. The body of the camera is made of steel or other metals or of plastic.

The Manufacturing Process

Making the Hardware Components

- Metal hardware components are manufactured from steel or steel and zinc alloys using precision metalworking equipment such as lathes and drill presses.
- If the external body shell of an inexpensive microscope is plastic, it is usually a light, rigid plastic such as acrylonitrile-butadiene-styrene (ABS) plastic. ABS plastic components are made by injection molding. In this process the plastic is melted and forced under pressure into a mold in the shape of the final product. The plastic is then allowed to cool back into a solid. The mold is opened and the product is removed.

Making Optical Glass

- The proper raw materials for the type of optical glass desired are mixed in the proper proportions, along with waste glass of the same type. This waste glass, known as cullet, acts as a flux. A flux is a substance which causes raw materials to react at a lower temperature than they would without it.
- The mixture is heated in a glass furnace until it has melted into a liquid. The temperature varies with the type of glass being made, but is typically about 2550 °F (1400 °C).
- The temperature is raised to about 2800 °F (1550 °C) to force air bubbles to rise to the surface. It is then slowly cooled and stirred constantly until it has reached a temperature of about 1800 °F (1000 °C). The glass is now an extremely thick liquid, which is poured into molds shaped like the lenses to be made.
- When the glass has cooled to about 600 °F (300 °C), it is reheated to about 1000 °F (500 °C). This process, known as annealing, removes internal stresses which form during the initial cooling period and which weaken the glass. The glass is then allowed to cool slowly to room temperature. The pieces of glass are removed from the molds. They are now known as blanks.

Making the Lenses

- The blank is now placed in a vise and held beneath a rapidly rotating cylindrical cutter with a diamond blade. This cutter, known as a curve generator, trims the surface of the blank until a close approximation of the desired curve is obtained. The cut lens is inspected and cut again if necessary. The difficulty of this process varies widely depending on the type of glass being cut and the exact curvature required. Several cuttings may be required, and the time involved may be a few minutes or more than half an hour.

- Several cut blanks are placed on the surface of a curved block in such a way that their curved surfaces line up as if they were all part of one spherical surface. This allows many lenses to be ground at the same time. A cast iron grinding surface known as a tool is placed on top on the lenses. The block of lenses rotates while the tool moves at random on top of it. A steady flow of liquid moves between the tool and the lenses. This liquid, known as a slurry, contains water, an abrasive (usually silicon carbide) to do the grinding, a coolant to prevent overheating, and a surfactant to keep the abrasive from settling out of the slurry. The lenses are inspected after grinding and reground if necessary. The grinding process may take one to eight hours.

- The lenses are moved to a polishing machine. This is similar to the grinding machine, but the tool is made of pitch (a thick, soft resin derived from tar). A pitch tool is made by placing tape around a curved dish, pouring in hot, liquid pitch, and letting it cool back into a solid. A pitch tool can be used about 50 times before it must be reshaped. It works in the same manner as a grinding tool, but instead of an abrasive the slurry contains a polishing substance (usually cerium dioxide). The lenses are inspected after polishing and the procedure is repeated as necessary. Polishing may take from half an hour to five hours. The lenses are cleaned and ready to be coated.

- The lenses are coated with magnesium fluoride. They are then inspected again, labeled with a date of manufacture and a serial number, and stored until needed.

Making the Mirror

- If a mirror is included, it is made in a way similar to the way in which a lens is made. Unlike a lens, it is cut, ground, and polished to be flat rather than curved. A reflective coating is then applied. Aluminum is heated in a vacuum to produce a vapor. A negative electrostatic charge is applied to the surface of the mirror so that it attracts the positively charged aluminum ions. This allows a thin, even coating of metal to be applied. A protective coating of silicon dioxide is then applied. Like a lens, the mirror is inspected, labeled, and stored.

Assembling the Microscope

- All of the final assembly of the microscope is done by hand. The workers wear gloves, masks, and gowns so that dirt does not damage the lenses or the internal mechanisms of the microscope. First the lenses are placed in the steel tubes, which make up the bodies of the eyepiece and the objective. These tubes are manufactured in standard sizes, which allow them to be assembled into a standard size microscope.

- The focusing mechanism of most microscopes is a rack and pinion system. This consists of a flat piece of metal with teeth on one side (the rack) and a metal wheel with teeth (the pinion), which controls the movement of the rack. The rack and pinion direct the objective so that its movement toward or away from the object being observed can be controlled. In many microscopes, the rack and pinion are attached to the stage (the flat metal plate on which the object being observed rests) and the objective remains stationary. After the rack and pinion system is installed, the knobs that control it are attached.
- The external body shell of the microscope is assembled around the internal focusing mechanism. The eyepiece (or two eyepieces, for a binocular microscope) and objective (or a rotating disk containing several different objectives) are screwed into place. Eyepieces and objectives are manufactured in standard sizes that allow many different eyepieces and objectives to be used in any standard microscope.
- If the microscope contains a mirror, this is attached to the body of the microscope below the opening in the stage. If it contains a light bulb instead, this may be attached in the same place (to shine light through the observed object) or it may be placed to the side of the stage (to shine light on top of the object). Some professional microscopes contain both kinds of light bulbs to allow both kinds of observation. If the microscope contains a camera, it is attached to the top of the body.
- The microscope is tested. If it functions correctly, the eyepiece and objective are usually unscrewed before packing. The parts of the microscope are packed securely in close-fitting compartments lined with cloth or foam. These compartments are often part of a wood or steel box. The microscope is then placed in a strong cardboard container and shipped to consumers.

Quality Control

The most critical part of quality control for a microscope is the accuracy of the lenses. During cutting and polishing, the size of the lens is measured with a vernier caliper. This device holds the lens between two jaws. One remains stationary while the other is gently moved into place until it touches the lens. The dimensions of the lens are read off a scale, which moves along with the movable jaw.

The curvature of the lens is measured with a spherometer. This device looks like a pocket watch with three small pins protruding from the base. The two outer pins remain in place, while the inner pin is allowed to move in or out. The movement of this pin is connected to a scale on the face of the spherometer. The scale reveals the degree of curvature of the lens. A typical lens should vary no more than about one-thousandth of an inch (25 micrometers).

During polishing, these tests are not accurate enough to ensure that the lens will focus light properly. Optical tests must be used. One typical test, known as an autocollimation test, involves shining a pinpoint light source through a lens in a dark room. A diffraction grating (a surface containing thousands of microscopic parallel grooves per inch) is placed at the point where the lens should focus the light. The grating causes a pattern of light and dark lines to form around the true focal point. It is compared with the theoretical focal point and the lens is repolished if necessary.

The mechanical parts of the microscope are also tested to ensure that they function correctly. The eyepiece and the objective must screw firmly into their proper places and must be perfectly centered to form a sharp image. The rack and pinion focusing mechanism is tested to ensure that it moves smoothly and that the distance between the objective and the stage is controlled precisely. Rotating disks containing multiple objectives are tested to be sure that they rotate smoothly and that each objective remains firmly in place during use.

BINOCULARS

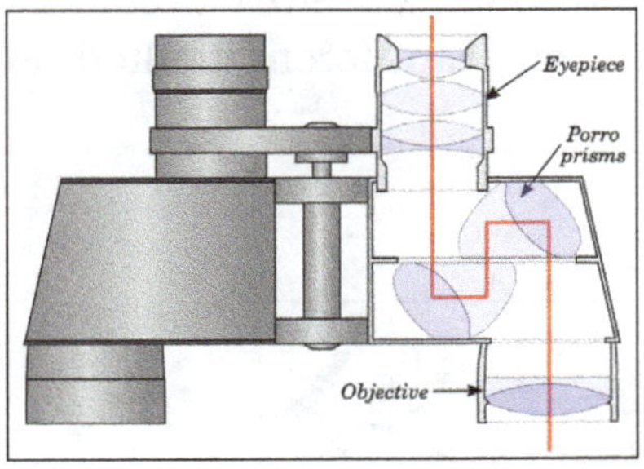

A typical Porro prism binoculars design.

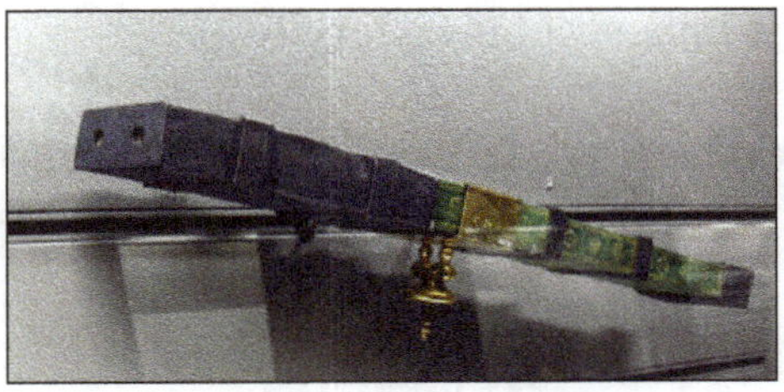
Binoculars.

Binoculars or field glasses are two telescopes mounted side-by-side and aligned to point in the same direction, allowing the viewer to use both eyes (binocular vision) when viewing distant objects. Most are sized to be held using both hands, although sizes vary widely from opera glasses to large pedestal mounted military models.

Unlike a (monocular) telescope, binoculars give users a three-dimensional (3D) image: for nearer objects the two views, presented to each of the viewer's eyes from slightly different viewpoints, produce a merged view with an impression of depth.

Optical Designs

Galilean Binoculars

Galilean Binoculars.

Almost from the invention of the telescope in the 17th century the advantages of mounting two of them side by side for binocular vision seems to have been explored. Most early binoculars used Galilean optics; that is, they used a convex objective and a concave eyepiece lens. The Galilean design has the advantage of presenting an erect image but has a narrow field of view and is not capable of very high magnification. This type of construction is still used in very cheap models and in opera glasses or theater glasses. The Galilean design is also used in low magnification binocular surgical and jewelers' loupes because they can be very short and produce an upright image without extra or unusual erecting optics, reducing expense and overall weight. They also have large exit pupils making centering less critical and the narrow field of view works well in those applications. These are typically mounted on an eyeglass frame or custom-fit onto eyeglasses.

Binoculars with Keplerian Optics

An improved image and higher magnification is achieved in binoculars employing Keplerian optics, where the image formed by the objective lens is viewed through a positive eyepiece lens (ocular). Since the Keplerian configuration produces an inverted image, different methods are used to turn the image right way up.

Binoculars with Erecting Lenses

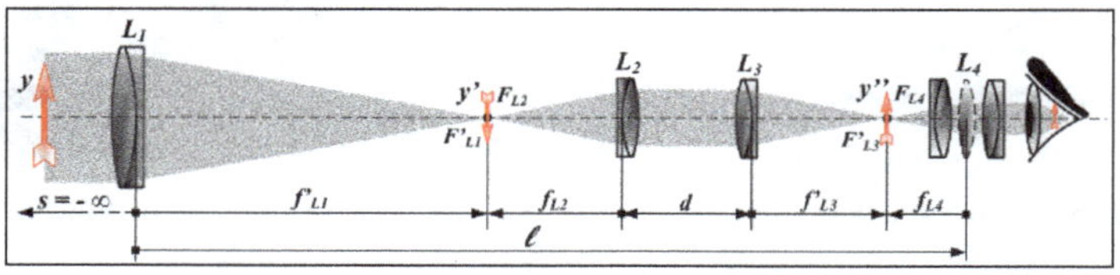

Cross-section of relay lens assembly – system.

In aprismatic binoculars with Keplerian optics (which were sometimes called "twin telescopes") each tube has one or two additional lenses (relay lens) between the objective and the ocular. These lenses are used to erect the image. The binoculars with erecting lenses had a serious disadvantage: they are too long. Such binoculars were popular in the 1800s, but became obsolete shortly after the Karl Zeiss company introduced improved prism binoculars in the 1890s.

Prism Binoculars

Optical prisms added to the design are another way to turn the image right way up, usually in a Porro prism or roof-prisms design.

Porro Prism Binoculars

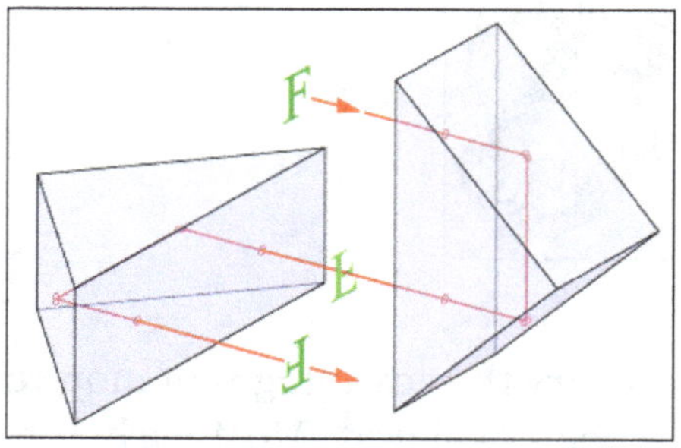

Double Porro prism design.

Porro prism binoculars.

Porro prism binoculars are named after Italian optician Ignazio Porro who patented this image erecting system in 1854, which was later refined by makers like the Carl Zeiss company in the 1890s. Binoculars of this type use a pair of Porro prisms in a Z-shaped configuration to erect the image. This results in binoculars that are wide, with objective lenses that are well separated and offset from the eyepieces, giving a better sensation of depth. Porro prism designs have the added benefit of folding the optical path so that the physical length of the binoculars is less than the focal length of the objective.

Roof-prisms Binoculars

Binoculars using roof prisms may have appeared as early as the 1870s in a design by Achille Victor Emile Daubresse. In 1897 Moritz Hensoldt began marketing roof prism binoculars. Most roof prism binoculars use either the Abbe-Koenig prism or the Schmidt-Pechan prism designs to erect the image and fold the optical path. They have objective lenses that are approximately in line with the eyepieces.

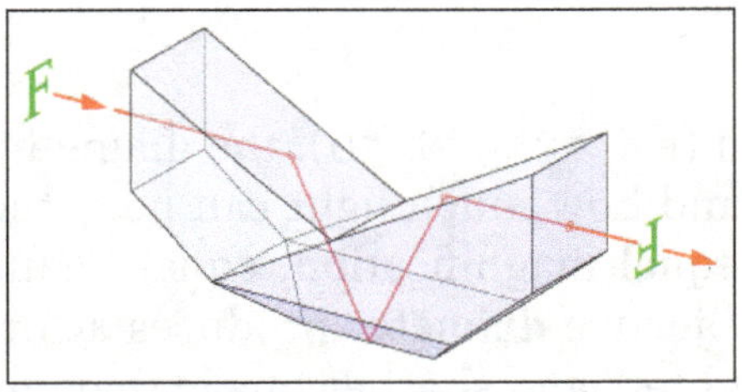

Abbe-Koenig "roof" prism design.

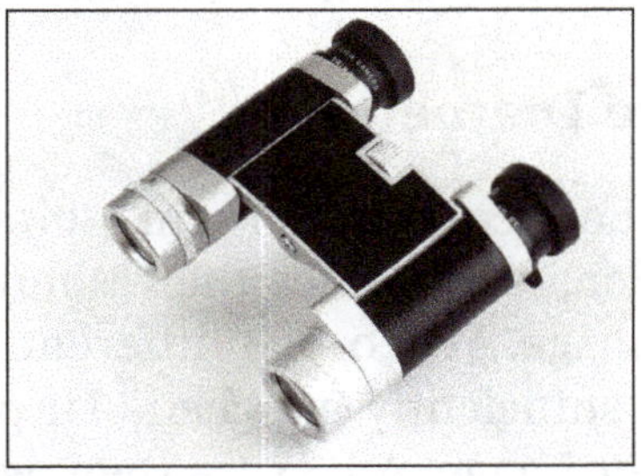

Binoculars with Schmidt-Pechan "roof" prisms.

Roof-prisms designs create an instrument that is narrower and more compact than Porro prisms. There is also a difference in image brightness. Porro-prism binoculars will inherently produce a brighter image than Schmidt-Pechan roof-prism binoculars of the same magnification, objective size, and optical quality, because this roof-prism design employs silvered surfaces that reduce light transmission by 12% to 15%. Roof-prisms designs also require tighter tolerances for alignment of their optical elements (collimation). This adds to their expense since the design requires them to use fixed elements that need to be set at a high degree of collimation at the factory. Porro prisms binoculars occasionally need their prism sets to be re-aligned to bring them into collimation. The fixed alignment in roof-prism designs means the binoculars normally will not need re-collimation.

Optical Parameters

Parameters listed on the prism cover plate describing 7 power magnification binoculars with a 50 mm objective diameter and a 372 foot (113 m) field of view at 1,000 yards (910 m).

Binoculars are usually designed for specific applications. These different designs require certain optical parameters which may be listed on the prism cover plate of the binoculars. Those parameters are.

Magnification

Given as the first number in a binocular description (e.g. 7×35, 8×50), magnification is the ratio of the focal length of the objective divided by the focal length of the eyepiece. This gives the magnifying power of binoculars (sometimes expressed as "diameters"). A magnification factor of 7, for example, produces an image 7 times larger than the original seen from that distance. The desirable amount of magnification depends upon the intended application, and in most binoculars is a permanent, non-adjustable feature of the device (zoom binoculars are the exception). Hand-held binoculars typically have magnifications ranging from 7x to 10x, so they will be less susceptible to the effects of shaking hands. A larger magnification leads to a smaller field of view and may require a tripod for image stability. Some specialized binoculars for astronomy or military use have magnifications ranging from 15x to 25x.

Objective Diameter

Given as the second number in a binocular description (e.g. 7×35, 8×50), the diameter of the objective lens determines the resolution (sharpness) and how much light can be gathered to form an image. When two different binoculars have equal magnification, equal quality, and produce a sufficiently matched exit pupil, the larger objective diameter produces a "brighter" and sharper image. An 8×40, then, will produce a "brighter" and sharper image than an 8×25, even though both enlarge the image an identical eight times. The larger front lenses in the 8×40 also produce wider beams of light (exit pupil) that leave the eyepieces. This makes it more comfortable to view with an 8×40 than an 8×25. A 10x50 binoculars is better than an 8x40 for magnification, sharpness and luminous flux. Objective diameter is usually expressed in millimeters. It is customary to categorize binoculars by the magnification × the objective diameter; e.g. 7×50. Smaller binoculars may have a diameter of as low as 22 mm; 35 mm and 50 mm is a common diameter for field binoculars; astronomical binoculars have diameters ranging from 70 mm to 150 mm.

Field of View

The field of view of a pair of binoculars depends on its optical design and in general is inversely proportional to the magnifying power. It is usually notated in a linear value, such as how many feet (meters) in width will be seen at 1,000 yards (or 1,000 m), or in an angular value of how many degrees can be viewed.

Exit Pupil

Binoculars concentrate the light gathered by the objective into a beam whose diameter, the exit pupil, is the objective diameter divided by the magnifying power. For maximum effective light-gathering and brightest image and to maximize the sharpness, the exit pupil should at least equal the diameter of the pupil of the human eye — about 7 mm at night and about 3 mm daytime, reducing with age. If the cone of light streaming out of the binoculars is *larger* than the pupil it is going into, any light larger than the pupil is wasted. In daytime use the human pupil is typically dilated about 3 mm, which is about the exit pupil of a 7×21 binocular. Much larger 7×50 binoculars will produce a cone of light bigger than the pupil it is entering, and this light will, in the daytime, be wasted. An exit pupil that is too *small* will also present an observer with a dimmer view since only a small portion of the light gathering surface of the retina is used. For applications where equipment has to be carried (birdwatching, hunting), users opt for much smaller (lighter) binoculars with an exit pupil that matches their expected iris diameter so they will have maximum resolution and are not carrying the weight of wasted aperture.

A larger exit pupil makes it easier to put the eye where it can receive the light: anywhere in the large exit pupil cone of light will do. This ease of placement helps avoid, especially in large field of view binoculars, vignetting, which brings to an image with the borders darkened because the light from them is partially blocked, and it means that the image can be quickly found which is important when looking at birds or game animals that move rapidly, or for a seaman on the deck of a pitching boat or ship. Narrow exit pupil binoculars may also be fatiguing because the instrument must be held exactly in place in front of the eyes to provide a useful image. Finally, many people use their binoculars at dusk, in overcast conditions, and at night, when their pupils are larger. Thus the daytime exit pupil is not a universally desirable standard. For comfort, ease of use, and flexibility in applications, larger binoculars with larger exit pupils are satisfactory choices even if their capability is not fully used by day.

Eye Relief

Eye relief is the distance from the rear eyepiece lens to the exit pupil or eye point. It is the distance the observer must position his or her eye behind the eyepiece in order to see an unvignetted image. The longer the focal length of the eyepiece, the greater the potential eye relief. Binoculars may have eye relief ranging from a few millimeters to 2.5 centimeters or more. Eye relief can be particularly important for eyeglass wearers. The eye of an eyeglass wearer is typically further from the eye piece which necessitates a longer eye relief in order to avoid vignetting and, in the extreme cases, to conserve the entire field of view. Binoculars with short eye relief can also be hard to use in instances where it is difficult to hold them steady.

Close Focus Distance

Close focus distance is the closest point that the binocular can focus on. This distance varies from about 0.5 m to 30 m, depending upon the design of the binoculars. If the close focus distance is short respect to the magnification, the binocular can be used also to see particulars not visible to the naked eye.

Eyepieces

Binocular eyepieces usually consist of three or more lens elements in two or more groups. The lens furthest from the viewer's eye is called the *field lens* and that closest to the eye the *eye lens*. The most common configuration is that invented in 1849 by Carl Kellner. In this arrangement, the eye lens is a plano-concave/double convex achromatic doublet (the flat part of the former facing the eye) and the field lens is a double-convex singlet. A reversed Kellner eyepiece was developed in 1975 and in it the field lens is a double concave/double convex achromatic doublet and the eye lens is a double convex singlet. The reverse Kellner provides 50% more eye relief and works better with small focal ratios as well as having a slightly wider field.

Wide field binoculars typically utilize some kind of Erfle configuration, patented in 1921. These have five or six elements in three groups. The groups may be two achromatic doublets with a double convex singlet between them or may all be achromatic doublets. These eyepieces tend not to perform as well as Kellner eyepieces at high power because they suffer from astigmatism and ghost images. However they have large eye lenses, excellent eye relief, and are comfortable to use at lower powers.

Mechanical Design

Focus and Adjustment

Central focusing binoculars with adjustable interpupillary distance.

Binoculars have a focusing arrangement which changes the distance between ocular and objective lenses. Normally there are two different arrangements used to provide focus, "independent focus" and "central focusing":

- *Independent focus* is an arrangement where the two telescopes are focused independently by adjusting each eyepiece. Binoculars designed for heavy field use, such as military applications, traditionally have used independent focusing.

- *Central focusing* is an arrangement which involves rotation of a central focusing wheel to adjust both tubes together. In addition, one of the two eyepieces can be further adjusted to compensate for differences between the viewer's eyes (usually by rotating the eyepiece in its mount). Because the focal change effected by the adjustable eyepiece can be measured in the customary unit of refractive power, the diopter, the adjustable eyepiece itself is often called a *diopter*. Once this adjustment has been made for a given viewer, the binoculars can be refocused on an object at a different distance by using the focusing wheel to adjust both tubes together without eyepiece readjustment.

There are "focus-free" or "fixed-focus" binoculars that have no focusing mechanism other than the eyepiece adjustments that are meant to be set for the user's eyes and left fixed. These are considered to be compromise designs, suited for convenience, but not well suited for work that falls outside their designed range.

Binoculars can be generally used without eyeglasses by myopic (near-sighted) or hyperopic (far-sighted) users simply by adjusting the focus a little further. Most manufacturers leave a little extra available focal-range beyond the infinity-stop/setting to account for this when focusing for infinity. People with severe astigmatism, however, may still need to use their glasses while using binoculars.

People using binoculars.

Some binoculars have adjustable magnification, *zoom binoculars*, intended to give the user the flexibility of having a single pair of binoculars with a wide range of magnifications, usually by moving a "zoom" lever. This is accomplished by a complex series of adjusting lenses similar to a zoom camera lens. These designs are noted to be a compromise and even a gimmick since they add bulk, complexity and fragility to the binocular. The complex optical path also leads to a narrow field of view and a large drop in brightness at high zoom. Models also have to match the magnification for both eyes throughout the zoom range and hold collimation to avoid eye strain and fatigue.

Most modern binoculars are also adjustable via a hinged construction that enables the distance between the two telescope halves to be adjusted to accommodate viewers with different eye separation or "interpupillary distance". Most are optimized for the interpupillary distance (typically 56 mm) for adults.

Image Stability

Some binoculars use image-stabilization technology to reduce shake at higher magnifications. This is done by having a gyroscope move part of the instrument, or by powered mechanisms driven by gyroscopic or inertial detectors, or via a mount designed to oppose and damp the effect of shaking movements. Stabilization may be enabled or disabled by the user as required. These techniques allow binoculars up to 20× to be hand-held, and much improve the image stability of lower-power instruments. There are some disadvantages: the image may not be quite as good as the best unstabilized binoculars when tripod-mounted, stabilized binoculars also tend to be more expensive and heavier than similarly specified non-stabilised binoculars.

Alignment

The two telescopes in binoculars are aligned in parallel (collimated), to produce a single circular, apparently three-dimensional, image. Misalignment will cause the binoculars to produce a double image. Even slight misalignment will cause vague discomfort and visual fatigue as the brain tries to combine the skewed images.

Alignment is performed by small movements to the prisms, by adjusting an internal support cell or by turning external set screws, or by adjusting the position of the objective via eccentric rings built into the objective cell. Alignment is usually done by a professional, although the externally mounted adjustment features can be accessed by the end user.

Optical Coatings

Binoculars with red-colored multicoatings.

Because a typical binocular has 6 to 10 optical elements with special characteristics and up to 16 air-to-glass surfaces, binocular manufacturers use different types of optical coatings for technical reasons and to improve the image they produce.

Anti-reflective Coatings

Anti-reflective coatings reduce light lost at every optical surface through reflection at each surface. Reducing reflection via anti-reflective coatings also reduces the amount of "lost" light present inside the binocular which would otherwise make the image appear hazy (low contrast). A pair of binoculars with good optical coatings may yield a brighter image than uncoated binoculars with a larger objective lens, on account of superior light transmission through the assembly. A classic lens-coating material is magnesium fluoride, which reduces reflected light from 5% to 1%. Modern lens coatings consist of complex multi-layers and reflect only 0.25% or less to yield an image with maximum brightness and natural colors.

Phase Correction Coatings

In binoculars with roof prisms the light path is split into two paths that reflect on either side of the roof prism ridge. One half of the light reflects from roof surface 1 to roof surface 2. The other half of the light reflects from roof surface 2 to roof surface 1. This causes the light to become partially polarized (due to a phenomenon called Brewster's angle). During subsequent reflections the direction of this polarization vector is changed but it is changed differently for each path in a manner similar to a Foucault pendulum. When the light following the two paths is recombined the polarization vectors of each path do not coincide. The angle between the two polarization vectors is called the *phase shift*, or the geometric phase, or the Berry phase. This interference between the two paths with different geometric phase results in a varying intensity distribution in the image reducing apparent contrast and resolution compared to a porro prism erecting system. These unwanted interference effects can be suppressed by vapor depositing a special dielectric coating known as a *phase-correction coating* or *P-coating* on the roof surfaces of the roof prism. This coating corrects for the difference in geometric phase between the two paths so both have effectively the same phase shift and no interference degrades the image.

Binoculars using either a Schmidt–Pechan roof prism or an Abbe–Koenig roof prism benefit from phase coatings. Porro prism binoculars do not split beams and therefore they do not require any phase coatings.

Metallic Mirror Coatings

In binoculars with Schmidt–Pechan roof prisms, mirror coatings are added to some surfaces of the roof prism because the light is incident at one of the prism's glass-air boundaries at an angle less than the critical angle so total internal reflection does not occur. Without a mirror coating most of that light would be lost. Schmidt–Pechan roof prism aluminum mirror coating (reflectivity of 87% to 93%) or silver mirror coating (reflectivity of 95% to 98%) is used.

In older designs silver mirror coatings were used but these coatings oxidized and lost reflectivity over time in unsealed binoculars. Aluminum mirror coatings were used in later unsealed designs because they did not tarnish even though they have a lower reflectivity than silver. Modern designs use either aluminum or silver. Silver is used in modern high-quality designs which are sealed and filled with a nitrogen or argon inert atmosphere so that the silver mirror coating does not tarnish.

Porro prism binoculars and roof prism binoculars using the Abbe–Koenig roof prism typically do not use mirror coatings because these prisms reflect with 100% reflectivity using total internal reflection in the prism.

Dielectric Mirror Coatings

Dielectric coatings are used in Schmidt–Pechan roof prisms to cause the prism surfaces to act as a dielectric mirror. The non-metallic dielectric reflective coating is formed from several multilayers of alternating high and low refractive index materials deposited on the roof prism's reflective surfaces. Each single multilayer reflects a narrow band of light frequencies so several multilayers, each tuned to a different color, are required to reflect white light. This multi-multilayer coating increases reflectivity from the prism surfaces by acting as a distributed Bragg reflector. A well-designed dielectric coating can provide a reflectivity of more than 99% across the visible light spectrum. This reflectivity is much improved compared to either an aluminium mirror coating (87% to 93%) or silver mirror coating (95% to 98%).

Porro prism binoculars and roof prism binoculars using the Abbe–Koenig roof prism do not use dielectric coatings because these prisms reflect with very high reflectivity using total internal reflection in the prism rather than requiring a mirror coating.

Terms used to Describe Coatings

Special reflective coatings on large naval binoculars.

For all Binoculars

The presence of any coatings is typically denoted on binoculars by the following terms:

- Coated optics: One or more surfaces are anti-reflective coated with a single-layer coating.
- Fully coated: All air-to-glass surfaces are anti-reflective coated with a single-layer coating. Plastic lenses, however, if used, may not be coated.
- Multi-coated: One or more surfaces have anti-reflective multi-layer coatings.
- Fully multi-coated: All air-to-glass surfaces are anti-reflective multi-layer coated.

For Binoculars with Roof Prisms Only

- Phase-coated or P-coating: The roof prism has a phase-correcting coating.
- Aluminium-coated: The roof prism mirrors are coated with an aluminium coating.
- Silver-coated: The roof prism mirrors are coated with a silver coating.
- Dielectric-coated: The roof prism mirrors are coated with a dielectric coating.

Applications

Use

Hand-held binoculars range from small 3 × 10 Galilean opera glasses, used in theaters, to glasses with 7 to 12 times magnification and 30 to 50 mm diameter objectives for typical outdoor use.

Tower Optical coin-operated binoculars.

Many tourist attractions have installed pedestal-mounted, coin-operated binoculars to allow visitors to obtain a closer view of the attraction.

Land Surveys and Geographic Data Collection

Although technology has surpassed using binoculars for data collection, historically these were advanced tools used by geographers and other geoscientists. Field glasses still today can provide visual aid when surveying large areas.

Bird Watching

Birdwatching is a very popular hobby among nature and animal lovers and a binocular is their most basic tool. Typically binoculars with a magnification of 7x to 10x are used.

Hunting

Hunters commonly use binoculars in the field as a way to see game animals that are too far away to spot with the naked eye. Hunters most commonly use 8x binoculars with light transmission and large enough objectives to gather light in low light conditions.

Range Finding

Many binoculars have a range finding reticle (scale) superimposed upon the view. This scale allows the distance to the object to be estimated if the object's height is known (or estimable). The common mariner 7×50 binoculars have these scales with the angle between marks equal to 5 mil. One mil is equivalent to the angle between the top and bottom of an object one meter in height at a distance of 1000 meters.

Therefore, to estimate the distance to an object that is a known height the formula is:

$$D = \frac{OH}{\text{Mil}} \times 1000$$

where:

- D is the *Distance* to the object in meters.
- OH is the known *Object Height*.
- Mil is the angular height of the object in number of *Mil*.

With the typical 5 mil scale (each mark is 5 mil), a lighthouse that is 3 marks high that is known to be 120 meters tall is 8000 meters distance.

$$8000\text{m} = \frac{120\text{m}}{15\text{mil}} \times 1000.$$

Military

Binoculars have a long history of military use. Galilean designs were widely used up to the end of the 19th century when they gave way to porro prism types. Binoculars constructed for general military use tend to be more rugged than their civilian counterparts. They generally avoid fragile center focus arrangements in favor of independent focus, which also makes for easier, more effective weatherproofing. Prism sets in military binoculars may have redundant aluminized coatings on their prism sets to guarantee they don't lose their reflective qualities if they get wet.

One variant form was called "trench binoculars", a combination of binoculars and periscope, often used for artillery spotting purposes. It projected only a few inches above the parapet, thus keeping the viewer's head safely in the trench.

Trench binoculars.

Military binoculars of the Cold War era were sometimes fitted with passive sensors that detected active IR emissions, while modern ones usually are fitted with filters blocking laser beams used as weapons. Further, binoculars designed for military usage may include a stadiametric reticle in one ocular in order to facilitate range estimation.

Naval ship binoculars.

There are binoculars designed specifically for civilian and military use at sea. Hand held models will be 5× to 7× but with very large prism sets combined with eyepieces designed to give generous eye relief. This optical combination prevents the image vignetting or going dark when the binoculars are pitching and vibrating relative to the viewer's eye. Large, high-magnification models with large objectives are also used in fixed mountings.

Very large binocular naval rangefinders (up to 15 meters separation of the two objective lenses, weight 10 tons, for ranging World War II naval gun targets 25 km away) have been used, although late-20th century technology made this application mostly redundant.

Astronomical

Binoculars are widely used by amateur astronomers; their wide field of view makes them useful for comet and supernova seeking (giant binoculars) and general observation (portable binoculars). Binoculars specifically geared towards astronomical viewing will have larger aperture objectives (in the 70 mm or 80 mm range) because the diameter of the objective lens increases the total amount of light captured, and therefore determines the faintest star that can be observed. Binoculars designed specifically for astronomical viewing (often 80 mm and larger) are sometimes designed without prisms in order to allow maximum light transmission. Such binoculars also usually have changeable eyepieces to vary magnification. Binoculars with high magnification and heavy weight usually require some sort of mount to stabilize the image. A magnification of 10x is generally considered the practical limit for observation with handheld binoculars. Binoculars more powerful than 15×70 require support of some type. Much larger binoculars have been made by amateur telescope makers, essentially using two refracting or reflecting astronomical telescopes.

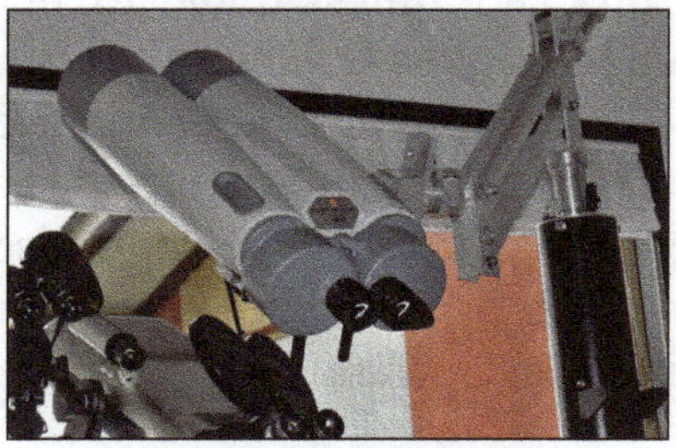

25 × 150 binoculars adapted for astronomical use.

Of particular relevance for low-light and astronomical viewing is the ratio between magnifying power and objective lens diameter. A lower magnification facilitates a larger field of view which is useful in viewing the Milky Way and large nebulous objects (referred to as deep sky objects) such as the nebulae and galaxies. The large (typical 7 mm using 7x50) exit pupil [objective (mm)/power] of these devices results in a small portion of the gathered light not being usable by individuals whose pupils do not sufficiently dilate. For example, the pupils of those over 50 rarely dilate over 5 mm wide. The large exit pupil also collects more light from the background sky, effectively decreasing contrast, making the detection of faint objects more difficult except perhaps in remote locations with negligible light pollution. Many astronomical objects of 8 magnitude or brighter, such as the star clusters, nebulae and galaxies listed in the Messier Catalog, are readily viewed in hand-held binoculars in the 35 to 40 mm range, as are found in many households for birding, hunting, and viewing sports events. For observing smaller star clusters, nebulae, and galaxies binocular magnification is an important factor for visibility because these objects appear tiny at typical binocular magnifications.

A simulated view of how the Andromeda Galaxy (Messier 31) would appear in a pair of binoculars.

Some open clusters, such as the bright double cluster (NGC 869 and NGC 884) in the constellation Perseus, and globular clusters, such as M13 in Hercules, are easy to spot. Among nebulae, M17 in Sagittarius and the North America Nebula (NGC 7000) in Cygnus are also readily viewed. Binoculars can show a few of the wider-split binary stars such as Albireo in the constellation Cygnus.

A number of solar system objects that are mostly to completely invisible to the human eye are reasonably detectable with medium-size binoculars, including larger craters on the Moon; the dim outer planets Uranus and Neptune; the inner "minor planets" Ceres, Vesta and Pallas; Saturn's largest moon Titan; and the Galilean moons of Jupiter. Although visible unaided in pollution-free skies, Uranus and Vesta require binoculars for easy detection. 10×50 binoculars are limited to an apparent magnitude of +9.5 to +11 depending on sky conditions and observer experience. Asteroids like Interamnia, Davida, Europa and, unless under exceptional conditions Hygiea, are too faint to be seen with commonly sold binoculars. Likewise too faint to be seen with most binoculars are the planetary moons except the Galileans and Titan, and the dwarf planets Pluto and Eris. Other difficult binocular targets include the phases of Venus and the rings of Saturn. Only binoculars with very high magnification, 20x or higher, are capable of discerning Saturn's rings to a recognizable extent. High-power binoculars can sometimes show one or two cloud belts on the disk of Jupiter if optics and observing conditions are sufficiently good.

CAMERA

A camera is an optical instrument to capture still images or to record moving images, which are stored in a physical medium such as in a digital system or on photographic film. A camera consists of a lens which focuses light from the scene, and a camera body which holds the image capture mechanism.

Leica Camera.

Digital Camera.

The still image camera is the main instrument in the art of photography and captured images may be reproduced later as a part of the process of photography, digital imaging, photographic printing. The similar artistic fields in the moving image camera domain are film, videography, and cinematography.

Sub-miniature spy camera.

The modern photographic camera evolved from the camera obscura. The functioning of the camera is very similar to the functioning of the human eye. The first permanent photograph was made in 1825 by Joseph Nicéphore Niépce.

Physics

A camera captures light photons, usually from the visible spectrum for human viewing, but in general could also be from other portions of the electromagnetic spectrum.

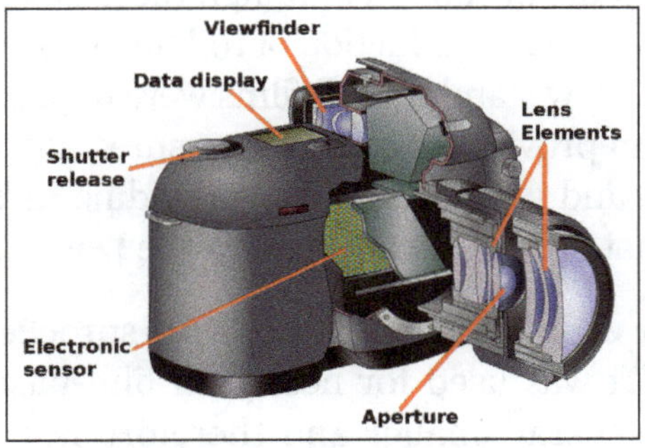

Basic elements of a modern still camera.

All cameras use the same basic design:

- Light enters an enclosed box through a converging or convex lens.
- An image is recorded on a light-sensitive medium (mainly a transition metal-halide).

A shutter mechanism controls the length of time that light can enter the camera.

Most photographic cameras have extra functions that allow:

- A person to view the scene to be recorded.
- A desired part of the scene to be in focus.
- Control of exposure so that it is not too bright or too dim.

On most digital cameras a display, often a liquid crystal display (LCD), permits the user to view the scene to be recorded and settings such as ISO speed, exposure, and shutter speed.

Still vs. Movie Cameras

A *still camera* is an optical device that creates a single image of an object or scene and records it on an electronic sensor or photographic film.

A movie camera or a video camera operates similarly to a still camera, except it records a series of static images in rapid succession, commonly at a rate of 24 frames per second. When the images are combined and displayed in order, the illusion of motion is achieved.

Cameras that capture many images in sequence are known as movie cameras or as ciné cameras in Europe; those designed for single images are still cameras. However these categories overlap as still cameras are often used to capture moving images in special effects work and many modern cameras can quickly switch between still and motion recording modes.

Image Capture on Film

Traditional cameras capture light onto photographic plate or photographic film. Video and digital cameras use an electronic image sensor, usually a charge coupled device (CCD) or a CMOS sensor to capture images which can be transferred or stored in a memory card or other storage inside the camera for later playback or processing.

A wide range of film and plate formats have been used by cameras. In the early history plate sizes were often specific for the make and model of camera although there quickly developed some standardisation for the more popular cameras. The introduction of roll film drove the standardization process still further so that by the 1950s only a few standard roll films were in use. These included 120 film providing 8, 12 or 16 exposures, 220 film providing 16 or 24 exposures, 127 film providing 8 or 12 exposures (principally in Brownie cameras) and 135 (35 mm film) providing 12, 20 or 36 exposures – or up to 72 exposures in the half-frame format or in bulk cassettes for the Leica Camera range.

For cine cameras, film 35 mm wide and perforated with sprocket holes was established as the standard format in the 1890s. It was used for nearly all film-based professional motion picture production. For amateur use, several smaller and therefore less expensive formats were introduced. 17.5 mm film, created by splitting 35 mm film, was one early amateur format, but 9.5 mm film, introduced in Europe in 1922, and 16 mm film, introduced in the US in 1923, soon became the standards for "home movies" in their respective hemispheres. In 1932, the even more economical 8 mm format was created by doubling the number of perforations in 16 mm film, then splitting it, usually after exposure and processing. The Super 8 format, still 8 mm wide but with smaller perforations to make room for substantially larger film frames, was introduced in 1965.

Film Speed

Traditionally used to "tell the camera" the film speed of the selected film on film cameras, film speed numbers are employed on modern digital cameras as an indication of the system's *gain* from light to numerical output and to control the automatic exposure system. Film speed is usually measured via the ISO system. The higher the film speed number the greater the film sensitivity to light, whereas with a lower number, the film is less sensitive to light.

White Balance

On digital cameras, electronic compensation for the color temperature associated with a given set of lighting conditions, ensuring that white light is registered as such on the imaging chip and therefore that the colors in the frame will appear natural. On mechanical, film-based cameras, this function is served by the operator's choice of film stock or with color correction filters. In addition to using white balance to register natural coloration of the image, photographers may employ white balance to aesthetic end, for example, white balancing to a blue object in order to obtain a warm color temperature.

Lens

The lens of a camera captures the light from the subject and brings it to a focus on the sensor. The design and manufacture of the lens is critical to the quality of the photograph being taken. The technological revolution in camera design in the 19th century revolutionized optical glass manufacture and lens design with great benefits for modern lens manufacture in a wide range of optical instruments from reading glasses to microscopes. Pioneers included Zeiss and Leitz.

Camera lenses are made in a wide range of focal lengths. They range from extreme wide angle, and standard, medium telephoto. Each lens is best suited to a certain type of photography. The extreme wide angle may be preferred for architecture because it has the capacity to capture a wide view of a building. The normal lens, because it often has a wide aperture, is often used for street and documentary photography. The telephoto lens is useful for sports and wildlife but it is more susceptible to camera shake.

Focus

Due to the optical properties of photographic lenses, only objects within a limited range of distances from the camera will be reproduced clearly. The process of adjusting this range is known as changing the camera's focus. There are various ways of focusing a camera accurately. The simplest cameras have fixed focus and use a small aperture and wide-angle lens to ensure that everything within a certain range of distance from the lens, usually around 3 metres (10 ft) to infinity, is in reasonable focus. Fixed focus cameras are usually inexpensive types, such as single-use cameras. The camera can also have a limited focusing range or scale-focus that is indicated on the camera body. The user will guess or calculate the distance to the subject and adjust the focus accordingly. On some cameras this is indicated by symbols (head-and-shoulders; two people standing upright; one tree; mountains).

The distance range in which objects appear clear and sharp, called depth of field, can be adjusted by many cameras. This allows for a photographer to control which objects appear in focus, and which do not.

Rangefinder cameras allow the distance to objects to be measured by means of a coupled parallax unit on top of the camera, allowing the focus to be set with accuracy. Single-lens reflex cameras allow the photographer to determine the focus and composition visually using the objective lens and a moving mirror to project the image onto a ground glass or plastic micro-prism screen. Twin-lens reflex cameras use an objective lens and a focusing lens unit (usually identical to the objective lens.) in a parallel body for composition and focusing. View cameras use a ground glass screen which is removed and replaced by either a photographic plate or a reusable holder containing sheet film before exposure. Modern cameras often offer autofocus systems to focus the camera automatically by a variety of methods.

Some experimental cameras, for example the planar Fourier capture array (PFCA), do not require focusing to allow them to take pictures. In conventional digital photography, lenses or mirrors map all of the light originating from a single point of an in-focus object to a single point at the sensor plane. Each pixel thus relates an independent piece of information about the far-away scene. In contrast, a PFCA does not have a lens or mirror, but each pixel has an idiosyncratic pair of diffraction gratings above it, allowing each pixel to likewise relate an independent piece of information (specifically, one component of the 2D Fourier transform) about the far-away scene. Together, complete scene information is captured and images can be reconstructed by computation.

Some cameras have post focusing. Post focusing means take the pictures first and then focusing later at the personal computer. The camera uses many tiny lenses on the sensor to capture light from every camera angle of a scene and is called plenoptics technology. A current plenoptic camera design has 40,000 lenses working together to grab the optimal picture.

Autofocus

On some cameras, the selection of a point in the imaging frame upon which the auto-focus system will attempt to focus. Many Single-lens reflex cameras (SLR) feature multiple auto-focus points in the viewfinder.

Exposure control settings on a digital camera.

Aperture

Adjustment of the lens opening measured as f-number, which controls the amount of light passing through the lens. Aperture also has an effect on depth of field and diffraction – the higher the f-number, the smaller the opening, the less light, the greater the depth of field, and the more the diffraction blur. The focal length divided by the f-number gives the effective aperture diameter.

The size of the aperture and the brightness of the scene controls the amount of light that enters the camera during a period of time, and the shutter controls the length of time that the light hits the recording surface. Equivalent exposures can be made using a large aperture size with a fast shutter speed and a small aperture with a slow shutter.

Shutter

Although a range of different shutter devices have been used during the development of the camera only two types have been widely used and remain in use today.

The Leaf shutter or more precisely the in-lens shutter is a shutter contained within the lens structure, often close to the diaphragm consisting of a number of metal leaves which are maintained under spring tension and which are opened and then closed when the shutter is released. The exposure time is determined by the interval between opening and closing. In this shutter design, the whole film frame is exposed at one time. This makes flash synchronisation much simpler as the flash only needs to fire once the shutter is fully open. Disadvantages of such shutters are their inability to reliably produce very fast shutter speeds (faster than 1/500th second or so) and the additional cost and weight of having to include a shutter mechanism for every lens.

The focal-plane shutter operates as close to the film plane as possible and consists of cloth curtains that are pulled across the film plane with a carefully determined gap between the two curtains (typically running horizontally) or consisting of a series of metal plates (typically moving vertically) just in front of the film plane. The focal-plane shutter is primarily associated with the single lens reflex type of cameras, since covering the film rather than blocking light passing through the lens allows the photographer to view through the lens at all times *except* during the exposure itself. Covering the film also facilitates removing the lens from a loaded camera (many SLRs have interchangeable lenses).

Adjustment of the speed (often expressed either as fractions of seconds or as an angle, with mechanical shutters) of the shutter to control the amount of time during which the imaging medium is exposed to light for each exposure. Shutter speed may be used to control the amount of light striking the image plane; 'faster' shutter speeds (that is, those of shorter duration) decrease both the amount of light and the amount of image blurring from motion of the subject or camera. The slower shutter speeds allow for long exposure shots that are done used to photograph images in very low light including the images of the night sky.

Metering

Measurement of exposure so that highlights and shadows are exposed according to the photographer's wishes. Many modern cameras meter and set exposure automatically. Before automatic exposure, correct exposure was accomplished with the use of a separate light metering device or by the photographer's knowledge and experience of gauging correct settings. To translate the amount of light into a usable aperture and shutter speed, the meter needs to adjust for the sensitivity of the film or sensor to light. This is done by setting the "film speed" or ISO sensitivity into the meter.

Exposure and Rendering

Camera controls are interrelated due to the similarities in the exposure controls. The total amount of light reaching the film plane (the 'exposure') changes with the duration of exposure, aperture of the lens, and on the effective focal length of the lens (which in variable focal length lenses, can force a change in aperture as the lens is zoomed). Changing any of these controls can alter the exposure. Many cameras may be set to adjust most or all of these controls automatically. This automatic functionality is useful for occasional photographers in many situations.

Manual shutter control and exposure settings can achieve unusual effects.

Star trails produced by long exposure photography in Chile. The prominent constellation in the middle is Orion, and to the right of it, the brightest star trail is Sirius.

The duration of an exposure is referred to as shutter speed, often even in cameras that do not have a physical shutter, and is typically measured in fractions of a second. It is quite possible to have exposures from one up to several seconds, usually for still-life subjects, and for night scenes exposure times can be several hours. However, longer shutter speeds blur motion, and shorter shutter speeds freeze motion. Therefore, moving subjects require fast shutter speeds.

The effective aperture is expressed by an f-number or f-stop (derived from *focal ratio*), which is proportional to the ratio of the focal length to the diameter of the aperture. Longer focal length lenses will pass less light through the same aperture diameter due to the greater distance the light has to travel; shorter focal length lenses will transmit more light through the same diameter of aperture.

The smaller the $f/$number, the larger the effective aperture. The present system of $f/$numbers to give the effective aperture of a lens was standardized by an international convention in 1963 and is referred to as the British Standard (BS-1013). Other aperture measurement scales had been used through the early 20th century, including the European Scale, Intermediate settings, and the 1881 Uniform System proposed by the Royal Photographic Society, which are all now largely obsolete.[:30] *T-stops* have been used for color motion picture lenses, to account for differences in *light transmission* through compound lenses, are calculated as $\text{T-number} = \text{f/ number x} \sqrt{\text{transmittance}}$.

If the f-number is decreased by a factor of √2, the aperture diameter is increased by the same factor, and its area is increased by a factor of 2. The f-stops that might be found on a typical lens include 2.8, 4, 5.6, 8, 11, 16, 22, 32, where going up "one stop" (using lower f-stop numbers) doubles the amount of light reaching the film, and stopping down one stop halves the amount of light.

Image capture can be achieved through various combinations of shutter speed, aperture, and film or sensor speed. Different (but related) settings of aperture and shutter speed enable photographs to be taken under various conditions of film or sensor speed, lighting and motion of subjects or camera, and desired depth of field. A slower speed film will exhibit less "grain", and a slower speed setting on an electronic sensor will exhibit less "noise", while higher film and sensor speeds allow for a faster shutter speed, which reduces motion blur or allows the use of a smaller aperture to increase the depth of field.

For example, a wider aperture is used for lower light and a lower aperture for more light. If a subject is in motion, then a high shutter speed may be needed. A tripod can also be helpful in that it enables a slower shutter speed to be used.

For example, *f*/8 at 8 ms (1/125 of a second) and *f*/5.6 at 4 ms (1/250 of a second) yield the same amount of light. The chosen combination affects the final result. The aperture and focal length of the lens determine the depth of field, which refers to the range of distances from the lens that will be in focus. A longer lens or a wider aperture will result in "shallow" depth of field (i.e., only a small plane of the image will be in sharp focus). This is often useful for isolating subjects from backgrounds as in individual portraits or macro photography.

Conversely, a shorter lens, or a smaller aperture, will result in more of the image being in focus. This is generally more desirable when photographing landscapes or groups of people. With very small apertures, such as pinholes, a wide range of distance can be brought into focus, but sharpness is severely degraded by diffraction with such small apertures. Generally, the highest degree of "sharpness" is achieved at an aperture near the middle of a lens's range (for example, *f*/8 for a lens with available apertures of *f*/2.8 to *f*/16). However, as lens technology improves, lenses are becoming capable of making increasingly sharp images at wider apertures.

A photographer using a tripod for greater stability during long exposure.

Image capture is only part of the image forming process. Regardless of material, some process must be employed to render the latent image captured by the camera into a viewable image. With slide film, the developed film is just mounted for projection. Print film requires the developed film negative to be printed onto photographic paper or transparency. Prior to the advent of laser jet and inkjet printers, celluloid photographic negative images had to be mounted in an enlarger which projected the image onto a sheet of light-sensitive paper for a certain length of time (usually measured in seconds or fractions of a second). This sheet then was soaked in a chemical bath of developer (to bring out the image) followed immediately by a stop bath (to neutralize the progression of development and prevent the image from changing further once exposed to normal light). After this, the paper was hung until dry enough to safely handle. This post-production process allowed the photographer to further manipulate the final image beyond what had already been captured on the negative, adjusting the length of time the image was projected by the enlarger and the duration of both chemical baths to change the image's intensity, darkness, clarity, etc. This process is still employed by both amateur and professional photographers, but the advent of digital imagery means that the vast majority of modern photographic work is captured digitally and rendered via printing processes that are no longer dependent on chemical reactions to light. Such digital images may be uploaded to an image server (e.g., a photo-sharing website), viewed on a television, or transferred to a computer or digital photo frame. Every type can then be produced as a hard copy on regular paper or photographic paper via a printer.

Prior to the rendering of a viewable image, modifications can be made using several controls. Many of these controls are similar to controls during image capture, while some are exclusive to the rendering process. Most printing controls have equivalent digital concepts, but some create different effects. For example, dodging and burning controls are different between digital and film processes.

Complexities

Professional medium format SLR (single-lens-reflex) cameras (typically using 120/220 roll film) use a hybrid solution, since such a large focal-plane shutter would be difficult to make and may run slowly. A manually inserted blade known as a dark slide allows the film to be covered when changing lenses or film backs. A blind inside the camera covers the film prior to and after the exposure (but is not designed to be able to give accurately controlled exposure times) and a leaf shutter that is normally *open* is installed in the lens. To take a picture, the leaf shutter closes, the blind opens, the leaf shutter opens then closes again, and finally the blind closes and the leaf shutter re-opens (the last step may only occur when the shutter is re-cocked).

Using a focal-plane shutter, exposing the whole film plane can take much longer than the exposure time. The exposure time does not depend on the time taken to make the exposure over all, only on the difference between the time a specific point on the film is uncovered and then covered up again. For example, an exposure of 1/1000 second may be achieved by the shutter curtains moving across the film plane in 1/50th of a second but with the two curtains only separated by 1/20th of the frame width. In fact in practice the curtains do not run at a constant speed as they would in an ideal design, obtaining an even exposure time depends mainly on being able to make the two curtains accelerate in a similar manner.

When photographing rapidly moving objects, the use of a focal-plane shutter can produce some unexpected effects, since the film closest to the start position of the curtains is exposed earlier than the film closest to the end position. Typically this can result in a moving object leaving a slanting image. The direction of the slant depends on the direction the shutter curtains run in (noting also that as in all cameras the image is inverted and reversed by the lens, i.e. "top-left" is at the bottom right of the sensor as seen by a photographer behind the camera).

Focal-plane shutters are also difficult to synchronise with flash bulbs and electronic flash and it is often only possible to use flash at shutter speeds where the curtain that opens to reveal the film completes its run and the film is fully uncovered, before the second curtain starts to travel and cover it up again. Typically 35mm film SLRs could sync flash at only up to 1/60th second if the camera has horizontal run cloth curtains, and 1/125th if using a vertical run metal shutter.

Camera Accessories

Accessories for cameras are mainly for care, protection, special effects and functions.

- Lens hood: Used on the end of a lens to block the sun or other light source to prevent glare and lens flare.
- Lens cap: Covers and protects the lens during storage.

- Lens adapter: Allows the use of lenses other than those for which the camera was designed.
- Lens filters: Allow artificial colors or change light density.
- Lens extension tubes allow close focus in macro photography.
- Flash equipment: Including light diffuser, mount and stand, reflector, soft box, trigger and cord.
- Care and protection: Including camera case and cover, maintenance tools, and screen protector.
- Camera monitor: Provides an off-camera view of the composition with a brighter and more colorful screen, and typically exposes more advanced tools such as framing guides, focus peaking, zebra stripes, waveform monitors (oftentimes as an "RGB parade"), vectorscopes and false color to highlight areas of the image critical to the photographer.
- Large format cameras use special equipment which includes magnifier loupe, view finder, angle finder, focusing rail /truck.
- Battery and sometimes a charger.
- Some professional SLR could be provided with interchangeable finders for eye-level or waist-level focusing, focusing screens, eye-cup, data backs, motor-drives for film transportation or external battery packs.
- Tripod, primarily used for keeping the camera steady while recording video, doing long exposure and timelapse photography.
- Microscope adapter, an adapter used to connect a camera to a microscope to photograph what the microscope is examining.
- Cable release, a remote shutter button that can be connected to the camera via a cable to remotely control the shutter, it can be used to lock the shutter open for a desired period of time. It is also commonly used to prevent camera shake from pressing the built in camera shutter button.
- Dew shield: Prevents moisture build up on the lens.
- UV filter, Can protect the front element of a lens from scratches, cracks, smudges, dirt, dust and moisture while keeping a minimum impact on image quality.

Camera Obscura

The forerunner to the photographic camera was the *camera obscura.* Camera obscura is the natural phenomenon that occurs when an image of a scene at the other side of a screen (or for instance a wall) is projected through a small hole in that screen and forms an inverted image (left to right and upside down) on a surface opposite to the opening.

The oldest known record of this principle is a description by Han Chinese philosopher Mozi. Mozi correctly asserted the camera obscura image is inverted because light travels inside the camera straight lines from its source.

Aristotle discovered that the smaller the hole, the clearer the image is when displayed. In addition, he is also known for the fact that when light passes through a small hole, the image being displayed will appear opposite on the wall or background.

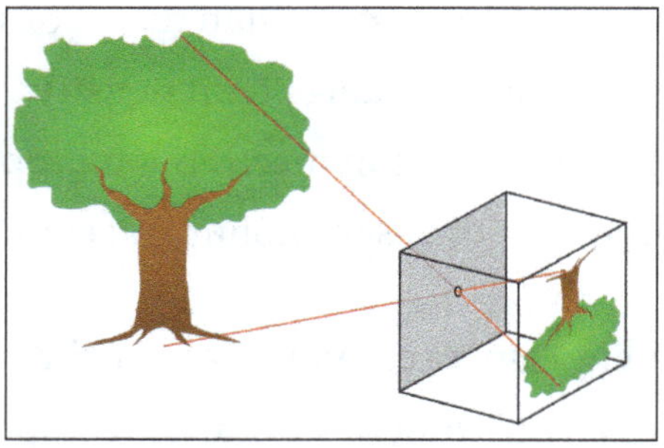

Camera obscura. Light enters a dark box through a small hole and creates an inverted image on the wall opposite the hole.

In the 11th century, Arab physicist Ibn al-Haytham (Alhazen) wrote very influential essays about the camera obscura, including experiments with light through a small opening in a darkened room. Ibn al-Haytam's writings on optics became very influential in Europe through Latin translations, inspiring people such as Witelo, John Peckham, Roger Bacon, Leonardo Da Vinci, René Descartes and Johannes Kepler.

Camera obscuras were used as drawing aids since at least circa 1550. Since the late 17th century, portable camera obscura devices in tents and boxes were used as drawing aids.

Photographic Camera

Before the development of the photographic camera, it had been known for hundreds of years that some substances, such as silver salts, darkened when exposed to sunlight. In a series of experiments, published in 1727, the German scientist Johann Heinrich Schulze demonstrated that the darkening of the salts was due to light alone, and not influenced by heat or exposure to air. The Swedish chemist Carl Wilhelm Scheele showed in 1777 that silver chloride was especially susceptible to darkening from light exposure, and that once darkened, it becomes insoluble in an ammonia solution. The first person to use this chemistry to create images was Thomas Wedgwood. To create images, Wedgwood placed items, such as leaves and insect wings, on ceramic pots coated with silver nitrate, and exposed the set-up to light. These images weren't permanent, however, as Wedgwood didn't employ a fixing mechanism. He ultimately failed at his goal of using the process to create fixed images created by a camera obscura.

View from the Window at Le Gras, the earliest surviving photograph.

The first permanent photograph of a camera image was made in 1825 by Joseph Nicéphore Niépce using a sliding wooden box camera made by Charles and Vincent Chevalier in Paris. Niépce had been experimenting with ways to fix the images of a camera obscura since 1816. The photograph

Niépce succeeded in creating shows the view from his window. It was made using an 8-hour exposure on pewter coated with bitumen. Niépce called his process "heliography". Niépce corresponded with the inventor Louis-Jacques-Mande Daguerre, and the pair entered into a partnership to improve the heliographic process. Niépce had experimented further with other chemicals, to improve contrast in his heliographs. Daguerre contributed an improved camera obscura design, but the partnership ended when Niépce died in 1833. Daguerre succeeded in developing a high-contrast and extremely sharp image by exposing on a plate coated with silver iodide, and exposing this plate again to mercury vapor. By 1837, he was able to fix the images with a common salt solution. He called this process *Daguerreotype*, and tried unsuccessfully for a couple years to commercialize it. Eventually, with help of the scientist and politician François Arago, the French government acquired Daguerre's process for public release. In exchange, pensions were provided to Daguerre as well as Niépce's son, Isidore.

In the 1830s, the English scientist Henry Fox Talbot independently invented a process to fix camera images using silver salts. Although dismayed that Daguerre had beaten him to the announcement of photography, on January 31, 1839 he submitted a pamphlet to the Royal Institution entitled *Some Account of the Art of Photogenic Drawing*, which was the first published description of photography. Within two years, Talbot developed a two-step process for creating photographs on paper, which he called calotypes. The calotyping process was the first to utilize negative prints, which reverse all values in the photograph – black shows up as white and vice versa. Negative prints allow, in principle, unlimited duplicates of the positive print to be made. Calotyping also introduced the ability for a printmaker to alter the resulting image through retouching. Calotypes were never as popular or widespread as daguerreotypes, owing mainly to the fact that the latter produced sharper details. However, because daguerreotypes only produce a direct positive print, no duplicates can be made. It is the two-step negative/positive process that formed the basis for modern photography.

The Giroux daguerreotype camera, the first to be commercially produced.

The first photographic camera developed for commercial manufacture was a daguerreotype camera, built by Alphonse Giroux in 1839. Giroux signed a contract with Daguerre and Isidore Niépce to produce the cameras in France, with each device and accessories costing 400 francs. The camera was a double-box design, with a landscape lens fitted to the outer box, and a holder for a ground glass focusing screen and image plate on the inner box. By sliding the inner box, objects at various distances could be brought to as sharp a focus as desired. After a satisfactory image had been focused on the screen, the screen was replaced with a sensitized plate. A knurled wheel controlled a copper flap in front of the lens, which functioned as a shutter. The early daguerreotype cameras required long exposure times, which in 1839 could be from 5 to 30 minutes.

After the introduction of the Giroux daguerreotype camera, other manufacturers quickly produced improved variations. Charles Chevalier, who had earlier provided Niépce with lenses, created in 1841 a double-box camera using a half-sized plate for imaging. Chevalier's camera had a hinged bed, allowing for half of the bed to fold onto the back of the nested box. In addition to having increased portability, the camera had a faster lens, bringing exposure times down to 3 minutes, and a prism at the front of the lens, which allowed the image to be laterally correct. Another French design emerged in 1841, created by Marc Antoine Gaudin. The Nouvel Appareil Gaudin camera had a metal disc with three differently-sized holes mounted on the front of the lens. Rotating to a different hole effectively provided variable f-stops, letting in different amount of light into the camera. Instead of using nested boxes to focus, the Gaudin camera used nested brass tubes. In Germany, Peter Friedrich Voigtländer designed an all-metal camera with a conical shape that produced circular pictures of about 3 inches in diameter. The distinguishing characteristic of the Voigtländer camera was its use of a lens designed by Joseph Petzval. The $f/3.5$ Petzval lens was nearly 30 times faster than any other lens of the period, and was the first to be made specifically for portraiture. Its design was the most widely used for portraits until Carl Zeiss introduced the anastigmat lens in 1889.

Within a decade of being introduced in America, 3 general forms of camera were in popular use: the American or chamfered-box camera, the Robert's-type camera or "Boston box", and the Lewis-type camera. The American-box camera had beveled edges at the front and rear, and an opening in the rear where the formed image could be viewed on ground glass. The top of the camera had hinged doors for placing photographic plates. Inside there was one available slot for distant objects, and another slot in the back for close-ups. The lens was focused either by sliding or with a rack and pinion mechanism. The Robert's-type cameras were similar to the American-box, except for having a knob-fronted worm gear on the front of the camera, which moved the back box for focusing. Many Robert's-type cameras allowed focusing directly on the lens mount. The third popular daguerreotype camera in America was the Lewis-type, introduced in 1851, which utilized a bellows for focusing. The main body of the Lewis-type camera was mounted on the front box, but the rear section was slotted into the bed for easy sliding. Once focused, a set screw was tightened to hold the rear section in place. Having the bellows in the middle of the body facilitated making a second, in-camera copy of the original image.

Daguerreotype cameras formed images on silvered copper plates and images were only able to develop with mercury vapor. The earliest daguerreotype cameras required several minutes to half an hour to expose images on the plates. By 1840, exposure times were reduced to just a few seconds owing to improvements in the chemical preparation and development processes, and to advances in lens design. American daguerreotypists introduced manufactured plates in mass production, and plate sizes became internationally standardized: whole plate (6.5 × 8.5 inches), three-quarter plate (5.5 × 7 1/8 inches), half plate (4.5 × 5.5 inches), quarter plate (3.25 × 4.25 inches), sixth plate (2.75 × 3.25 inches), and ninth plate (2 × 2.5 inches). Plates were often cut to fit cases and jewelry with circular and oval shapes. Larger plates were produced, with sizes such as 9 × 13 inches ("double-whole" plate), or 13.5 × 16.5 inches.

The collodion wet plate process that gradually replaced the daguerreotype during the 1850s required photographers to coat and sensitize thin glass or iron plates shortly before use and expose them in the camera while still wet. Early wet plate cameras were very simple and little different from Daguerreotype cameras, but more sophisticated designs eventually appeared. The Dubroni of 1864 allowed the sensitizing and developing of the plates to be carried out inside the camera itself rather than in a separate darkroom. Other cameras were fitted with multiple lenses for photographing several small portraits on a single larger plate, useful when making cartes de visite. It was during the wet plate era that the use of bellows for focusing became widespread, making the bulkier and less easily adjusted nested box design obsolete.

For many years, exposure times were long enough that the photographer simply removed the lens cap, counted off the number of seconds (or minutes) estimated to be required by the lighting conditions, then replaced the cap. As more sensitive photographic materials became available, cameras began to incorporate mechanical shutter mechanisms that allowed very short and accurately timed exposures to be made.

The use of photographic film was pioneered by George Eastman, who started manufacturing paper film in 1885 before switching to celluloid in 1889. His first camera, which he called the "Kodak", was first offered for sale in 1888. It was a very simple box camera with a fixed-focus lens and single shutter speed, which along with its relatively low price appealed to the average consumer. The Kodak came pre-loaded with enough film for 100 exposures and needed to be sent back to the factory for processing and reloading when the roll was finished. By the end of the 19th century Eastman had expanded his lineup to several models including both box and folding cameras.

Films also made possible capture of motion (cinematography) establishing the movie industry by end of 19th century.

Single Lens Reflex Camera

In photography, the single-lens reflex camera (SLR) is provided with a mirror to redirect light from the picture taking lens to the viewfinder prior to releasing the shutter for composing and focusing an image. When the shutter is released, the mirror swings up and away allowing the exposure of the photographic medium and instantly returns after the exposure. No SLR camera before 1954 had this feature, although the mirror on some early SLR cameras was entirely operated by the force exerted on the shutter release and only returned when the finger pressure was released. The Asahiflex II, released by Japanese company Asahi (Pentax) in 1954, was the world's first SLR camera with an instant return mirror.

Plate Camera

The earliest cameras produced in significant numbers used sensitised glass plates were *plate cameras.* Light entered a lens mounted on a lens board which was separated from the plate by an extendible bellows.There were simple box cameras for glass plates but also single-lens reflex cameras with interchangeable lenses and even for color photography (Autochrome Lumière). Many of these cameras had controls to raise or lower the lens and to tilt it forwards or backwards to control perspective.

Focusing of these plate cameras was by the use of a ground glass screen at the point of focus. Because lens design only allowed rather small aperture lenses, the image on the ground glass screen was faint and most photographers had a dark cloth to cover their heads to allow focussing and composition to be carried out more easily. When focus and composition were satisfactory, the ground glass screen was removed and a sensitised plate put in its place protected by a dark slide. To make the exposure, the dark slide was carefully slid out and the shutter opened and then closed and the dark slide replaced.

Glass plates were later replaced by sheet film in a dark slide for sheet film; adaptor sleeves were made to allow sheet film to be used in plate holders. In addition to the ground glass, a simple optical viewfinder was often fitted. Cameras which take single exposures on sheet film and are functionally identical to plate cameras were used for static, high-image-quality work; much longer in 20th century.

Folding Camera

The introduction of films enabled the existing designs for plate cameras to be made much smaller and for the base-plate to be hinged so that it could be folded up compressing the bellows. These designs were very compact and small models were dubbed *vest pocket* cameras. Folding rollfilm cameras were preceded by folding plate cameras, more compact than other designs.

Box Camera

Box cameras were introduced as a budget level camera and had few if any controls. The original box Brownie models had a small reflex viewfinder mounted on the top of the camera and had no aperture or focusing controls and just a simple shutter. Later models such as the Brownie 127 had larger direct view optical viewfinders together with a curved film path to reduce the impact of deficiencies in the lens.

Rangefinder Camera

As camera a lens technology developed and wide aperture lenses became more common, rangefinder cameras were introduced to make focusing more precise. Early rangefinders had two separate viewfinder windows, one of which is linked to the focusing mechanisms and moved right or left as the focusing ring is turned. The two separate images are brought together on a ground glass viewing screen. When vertical lines in the object being photographed meet exactly in the combined image, the object is in focus. A normal composition viewfinder is also provided. Later the viewfinder and rangefinder were combined. Many rangefinder cameras had interchangeable lenses, each lens requiring its own range- and viewfinder linkages.

Rangefinder cameras were produced in half- and full-frame 35 mm and rollfilm.

Instant Camera

After exposure every photograph is taken through pinch rollers inside of the instant camera. Thereby the developer paste contained in the paper 'sandwich' distributes on the image. After a minute, the cover sheet just needs to be removed and one gets a single original positive image with a fixed format. With some systems it was also possible to create an instant image negative, from which then could be made copies in the photo lab. The ultimate development was the SX-70 system of Polaroid, in which a row of ten shots – engine driven – could be made without having to remove any cover sheets from the picture. There were instant cameras for a variety of formats, as well as adapters for instant film use in medium- and large-format cameras.

Single-lens Reflex Camera

In the single-lens reflex camera, the photographer sees the scene through the camera lens. This avoids the problem of parallax which occurs when the viewfinder or viewing lens is separated from the taking lens. Single-lens reflex cameras have been made in several formats including sheet film 5×7" and 4×5", roll film 220/120 taking 8, 10, 12 or 16 photographs on a 120 roll and twice that number of a 220 film. These correspond to 6×9, 6×7, 6×6 and 6×4.5 respectively (all dimensions in cm). Notable manufacturers of large format and roll film SLR cameras include Bronica, Graflex, Hasselblad, Mamiya, and Pentax. However the most common format of SLR cameras has been 35 mm and subsequently the migration to digital SLR cameras, using almost identical sized bodies and sometimes using the same lens systems.

Almost all SLR cameras use a front surfaced mirror in the optical path to direct the light from the lens via a viewing screen and pentaprism to the eyepiece. At the time of exposure the mirror is flipped up out of the light path before the shutter opens. Some early cameras experimented with other methods of providing through-the-lens viewing, including the use of a semi-transparent pellicle as in the Canon *Pellix* and others with a small periscope such as in the Corfield Periflex series.

Twin-lens Reflex Camera

Twin-lens reflex cameras used a pair of nearly identical lenses, one to form the image and one as a viewfinder. The lenses were arranged with the viewing lens immediately above the taking lens. The viewing lens projects an image onto a viewing screen which can be seen from above. Some manufacturers such as Mamiya also provided a reflex head to attach to the viewing screen to allow the camera to be held to the eye when in use. The advantage of a TLR was that it could be easily focussed using the viewing screen and that under most circumstances the view seen in the viewing screen was identical to that recorded on film. At close distances however, parallax errors were encountered and some cameras also included an indicator to show what part of the composition would be excluded.

Some TLR had interchangeable lenses but as these had to be paired lenses they were relatively heavy and did not provide the range of focal lengths that the SLR could support. Most TLRs used 120 or 220 film; some used the smaller 127 film.

Large-format Camera

The large-format camera, taking sheet film, is a direct successor of the early plate cameras and remained in use for high quality photography and for technical, architectural and industrial photography. There are three common types, the view camera with its monorail and field camera variants, and the press camera. They have an extensible bellows with the lens and shutter mounted on a lens plate at the front. Backs taking rollfilm, and later digital backs are available in addition to the standard dark slide back. These cameras have a wide range of movements allowing very close control of focus and perspective. Composition and focusing is done on view cameras by viewing a ground-glass screen which is replaced by the film to make the exposure; they are suitable for static subjects only, and are slow to use.

Medium-format Camera

Medium-format cameras have a film size between the large-format cameras and smaller 35 mm cameras. Typically these systems use 120 or 220 rollfilm. The most common image sizes are 6×4.5 cm, 6×6 cm and 6×7 cm; the older 6×9 cm is rarely used. The designs of this kind of camera show greater variation than their larger brethren, ranging from monorail systems through the classic Hasselblad model with separate backs, to smaller rangefinder cameras. There are even compact amateur cameras available in this format.

Subminiature Camera

Cameras taking film significantly smaller than 35 mm were made. Subminiature cameras were first produced in the nineteenth century. The expensive 8×11 mm Minox, the only type of camera produced by the company from 1937 to 1976, became very widely known and was often used for espionage (the Minox company later also produced larger cameras). Later inexpensive subminiatures were made for general use, some using rewound 16 mm cine film. Image quality with these small film sizes was limited.

Movie Camera

A ciné camera or movie camera takes a rapid sequence of photographs on image sensor or strips of film. In contrast to a still camera, which captures a single snapshot at a time, the ciné camera takes a series of images, each called a "frame" through the use of an intermittent mechanism.

The frames are later played back in a ciné projector at a specific speed, called the "frame rate" (number of frames per second). While viewing, a person's eyes and brain merge the separate pictures to create the illusion of motion. The first ciné camera was built around 1888 and by 1890 several types were being manufactured. The standard film size for ciné cameras was quickly established as 35mm film and this remained in use until transition to digital cinematography. Other professional standard formats include 70 mm film and 16mm film whilst amateurs film makers used 9.5 mm film, 8mm film or Standard 8 and Super 8 before the move into digital format.

The size and complexity of ciné cameras varies greatly depending on the uses required of the camera. Some professional equipment is very large and too heavy to be hand held whilst some amateur cameras were designed to be very small and light for single-handed operation.

Camcorders

A camcorder is an electronic device combining a video camera and a video recorder. Although marketing materials may use the colloquial term "camcorder", the name on the package and manual is often "video camera recorder". Most devices capable of recording video are camera phones and digital cameras primarily intended for still pictures; the term "camcorder" is used to describe a portable, self-contained device, with video capture and recording its primary function.

Professional Video Camera

A professional video camera (often called a television camera even though the use has spread beyond television) is a high-end device for creating electronic moving images (as opposed to a movie camera, that earlier recorded the images on film). Originally developed for use in television studios, they are now also used for music videos, direct-to-video movies, corporate and educational videos, marriage videos etc.

These cameras earlier used vacuum tubes and later electronic sensors.

Digital Camera

A digital camera (or digicam) is a camera that encodes digital images and videos digitally and stores them for later reproduction. Most cameras sold today are digital, and digital cameras are incorporated into many devices ranging from mobile phones (called camera phones) to vehicles.

Digital and film cameras share an optical system, typically using a lens with a variable diaphragm to focus light onto an image pickup device. The diaphragm and shutter admit the correct amount of light to the imager, just as with film but the image pickup device is electronic rather than chemical. However, unlike film cameras, digital cameras can display images on a screen immediately after being recorded, and store and delete images from memory. Most digital cameras can also record moving videos with sound. Some digital cameras can crop and stitch pictures and perform other elementary image editing.

Consumers adopted digital cameras in 1990s. Professional video cameras transitioned to digital around the 2000s–2010s. Finally movie cameras transitioned to digital in the 2010s.

The first camera using digital electronics to capture and store images was developed by Kodak engineer Steven Sasson in 1975. He used a charge-coupled device (CCD) provided by Fairchild Semiconductor, which provided only 0.01 megapixels to capture images. Sasson combined the CCD device with movie camera parts to create a digital camera that saved black and white images onto a cassette tape. The images were then read from the cassette and viewed on a TV monitor. Later, cassette tapes were replaced by flash memory.

In 1986, Japanese company Nikon introduced an analog-recording electronic single-lens reflex camera, the Nikon SVC.

The first full-frame digital SLR cameras were developed in Japan from around 2000 to 2002: the MZ-D by Pentax, the N Digital by Contax's Japanese R6D team, and the EOS-1Ds by Canon. Gradually in the 2000s, the full-frame DSLR became the dominant camera type for professional photography.

Camera Phone

In 2000, Sharp introduced the world's first digital camera phone, the J-SH04 J-Phone, in Japan. By the mid-2000s, higher-end cell phones had an integrated digital camera. By the beginning of the 2010s, almost all smartphones had an integrated digital camera.

Panoramic Camera

Panoramic cameras are fixed-lens digital action cameras. They usually have a single fish-eye lens or multiple lenses, to cover the entire 180° up to 360° in their field of view.

VR Camera

VR Cameras are panoramic cameras that also cover the top and bottom in their field of view. There have also been camera rigs employing multiple cameras to cover the whole 360° by 360° field of view.

OPTICAL INTERFEROMETER

Optical interferometer is an instrument used for making precise measurements for beams of light of such factors as length, surface irregularities, and index of refraction. It divides a beam of light into a number of beams that travel unequal paths and whose intensities, when reunited, add or subtract (interfere with each other). This interference appears as a pattern of light and dark bands called interference fringes. Information derived from fringe measurements is used for precise wavelength determinations, measurement of very small distances and thicknesses, the study of spectrum lines, and determination of refractive indices of transparent materials. In astronomy, interferometers are used to measure the distances between stars and the diameters of stars.

In 1881 the American physicist A.A. Michelson constructed the interferometer used in the Michelson-Morley experiment. The Michelson interferometer and its modifications are used in the optical industry for testing lenses and prisms, for measuring index of refraction, and for examining minute details of surfaces (microtopographies). The instrument consists of a half-silvered mirror that divides a light beam into two equal parts, one of which is transmitted to a fixed mirror and the other of which is reflected to a movable mirror. By counting the fringes created as the mirror is moved, the amount of movement can be precisely determined. Michelson also developed the stellar interferometer, capable of measuring the diameters of stars in terms of the angle, as small as 0.01″ of an arc, subtended by the extreme points of the star at the point of observation.

In 1896 the British physicist Lord Rayleigh described the Rayleigh interference refractometer, still widely used for determining the refractive indices of gases and liquids. It is a split-beam instrument, like the Michelson interferometer. One beam serves as a reference, while the other is passed first through a material of known index of refraction and then through the unknown. The index of refraction of the unknown can be determined by the displacement of its interference fringes from those of the known material.

The Fabry-Pérot interferometer (variable-gap interferometer) was produced in 1897 by the French physicists Charles Fabry and Alfred Pérot. It consists of two highly reflective and strictly parallel plates called an etalon. Because of the high reflectivity of the plates of the etalon, the successive multiple reflections of light waves diminish very slowly in intensity and form very narrow, sharp fringes. These may be used to reveal hyperfine structures in line spectra, to evaluate the widths of narrow spectral lines, and to redetermine the length of the standard metre.

The Fizeau-Laurent surface interferometer reveals departures of polished surfaces from a plane. The system was described by the French physicist A.-H.-L. Fizeau in 1862 and adapted in 1883 into the instruments now widely used in the optical industry. In the Fizeau-Laurent system, monochromatic light (light of a single colour) is passed through a pinhole and illuminates a reference plane and a workpiece directly below it. The light beam is perpendicular to the workpiece. By maintaining a slight angle between the surface of the workpiece and the surface of the plane of reference, fringes of equal thickness can be seen through a reflector placed above them. The fringes constitute a contour map of the surface of the workpiece, enabling an optical polisher to see and to remove defects and departures from flatness.

The Twyman-Green interferometer, an adaptation of the Michelson instrument introduced in 1916 by the English electrical engineer Frank Twyman and the English chemist Arthur Green, is used for testing lenses and prisms. It uses a point source of monochromatic light at the focus of a quality lens. When the light is directed toward a perfect prism, it returns to a viewing point exactly as it was from the source, and a uniform field of illumination is seen. Local imperfections in the prism glass distort the wave front. When the light is directed toward a lens backed by a convex mirror, it passes through the lens, strikes the mirror, and retraces its path through the lens to a viewing point. Imperfections in the lens result in fringe distortions.

PHOTOMETER

A photometer is an instrument that measures the strength of electromagnetic radiation in the range from ultraviolet to infrared and including the visible spectrum. Most photometers convert light into an electric current using a photoresistor, photodiode, or photomultiplier.

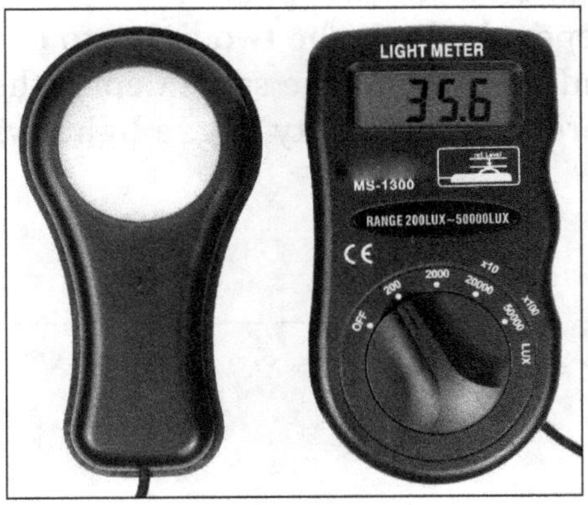

A photometer.

Photometers measure:

- Illuminance
- Irradiance
- Light absorption
- Scattering of light
- Reflection of light
- Fluorescence
- Phosphorescence
- Luminescence

Before electronic light sensitive elements were developed, photometry was done by estimation by the eye. The relative luminous flux of a source was compared with a standard source. The photometer is placed such that the illuminance from the source being investigated is equal to the standard source, as the human eye can judge equal illuminance. The relative luminous fluxes can then be calculated as the illuminance decreases proportionally to the inverse square of distance. A standard example of such a photometer consists of a piece of paper with an oil spot on it that makes the paper slightly more transparent. When the spot is not visible from either side, the illuminance from the two sides is equal.

By 1861, three types were in common use. These were Rumford's photometer, Ritchie's photometer, and photometers that used the extinction of shadows, which was considered to be the most precise.

Rumford's Photometer

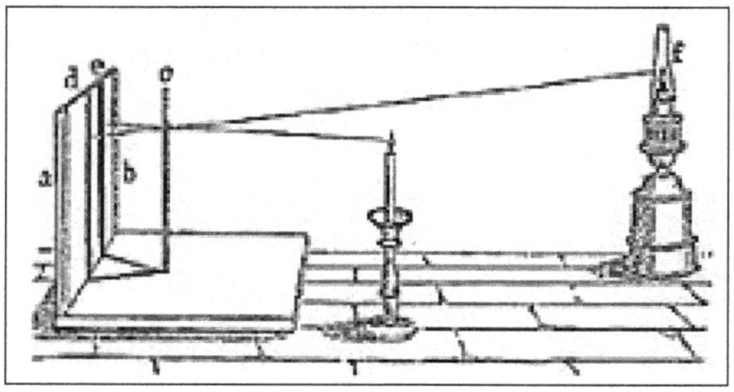

Rumford's photometer.

Rumford's photometer (also called a shadow photometer) depended on the principle that a brighter light would cast a deeper shadow. The two lights to be compared were used to cast a shadow onto paper. If the shadows were of the same depth, the difference in distance of the lights would indicate the difference in intensity (e.g. a light twice as far would be four times the intensity).

Ritchie's Photometer

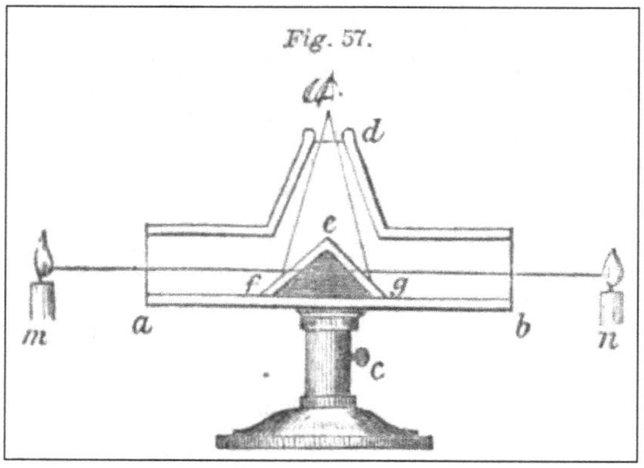

Ritchie's photometer.

Ritchie's photometer depends upon equal illumination of surfaces. It consists of a box (a,b) six or eight inches long, and one in width and depth. In the middle, a wedge of wood (f,e,g) was angled upwards and covered with white paper. The user's eye looked through a tube (d) at the top of a box. The height of the apparatus was also adjustable via the stand (c). The lights to compare were placed at the side of the box (m, n)—which illuminated the paper surfaces so that the eye saw both surfaces at once. By changing the position of the lights, they were made to illuminate both surfaces equally, with the difference in intensity corresponding to the square of the difference in distance.

Method of Extinction of Shadows

This type of photometer depended on the fact that if a light throws the shadow of an opaque object onto a white screen, there is a certain distance that, if a second light is brought there, obliterates all traces of the shadow.

Principle of Photometers

Most photometers detect the light with photoresistors, photodiodes or photomultipliers. To analyze the light, the photometer may measure the light after it has passed through a filter or through a monochromator for determination at defined wavelengths or for analysis of the spectral distribution of the light.

Photon Counting

Some photometers measure light by counting individual photons rather than incoming flux. The operating principles are the same but the results are given in units such as photons/cm^2 or photons·cm^{-2}·sr^{-1} rather than W/cm^2 or W·cm^{-2}·sr^{-1}.

Due to their individual photon counting nature, these instruments are limited to observations where the irradiance is low. The irradiance is limited by the time resolution of its associated detector readout electronics. With current technology this is in the megahertz range. The maximum irradiance is also limited by the throughput and gain parameters of the detector itself.

The light sensing element in photon counting devices in NIR, visible and ultraviolet wavelengths is a photomultiplier to achieve sufficient sensitivity.

In airborne and space-based remote sensing such photon counters are used at the upper reaches of the electromagnetic spectrum such as the X-ray to far ultraviolet. This is usually due to the lower radiant intensity of the objects being measured as well as the difficulty of measuring light at higher energies using its particle-like nature as compared to the wavelike nature of light at lower frequencies. Conversely, radiometers are typically used for remote sensing from the visible, infrared though radio frequency range.

Photography

Photometers are used to determine the correct exposure in photography. In modern cameras, the photometer is usually built in. As the illumination of different parts of the picture varies, advanced photometers measure the light intensity in different parts of the potential picture and use an algorithm to determine the most suitable exposure for the final picture, adapting the algorithm to the type of picture intended. Historically, a photometer was separate from the camera and known as an exposure meter. The advanced photometers then could be used either to measure the light from the potential picture as a whole, to measure from elements of the picture to ascertain that the most important parts of the picture are optimally exposed, or to measure the incident light to the scene with an integrating adapter.

Visible Light Reflectance Photometry

A reflectance photometer measures the reflectance of a surface as a function of wavelength. The surface is illuminated with white light, and the reflected light is measured after passing through a monochromator. This type of measurement has mainly practical applications, for instance in the paint industry to characterize the colour of a surface objectively.

UV and Visible Light Transmission Photometry

These are optical instruments for measurement of the absorption of light of a given wavelength (or a given range of wavelengths) of coloured substances in solution. From the light absorption, Beer's law makes it possible to calculate the concentration of the coloured substance in the solution. Due to its wide range of application and its reliability and robustness, the photometer has become one of the principal instruments in biochemistry and analytical chemistry. Absorption photometers for work in aqueous solution work in the ultraviolet and visible ranges, from wavelength around 240 nm up to 750 nm.

The principle of spectrophotometers and filter photometers is that (as far as possible) monochromatic light is allowed to pass through a container (cell) with optically flat windows containing the solution. It then reaches a light detector, that measures the intensity of the light compared to the intensity after passing through an identical cell with the same solvent but without the coloured substance. From the ratio between the light intensities, knowing the capacity of the coloured substance to absorb light (the absorbency of the coloured substance, or the photon cross section area of the molecules of the coloured substance at a given wavelength), it is possible to calculate the concentration of the substance using Beer's law.

Two types of photometers are used: spectrophotometer and filter photometer. In spectrophotometers a monochromator (with prism or with grating) is used to obtain monochromatic light of one defined wavelength. In filter photometers, optical filters are used to give the monochromatic light. Spectrophotometers can thus easily be set to measure the absorbance at different wavelengths, and they can also be used to scan the spectrum of the absorbing substance. They are in this way more flexible than filter photometers, also give a higher optical purity of the analyzing light, and therefore they are preferably used for research purposes. Filter photometers are cheaper, robuster and easier to use and therefore they are used for routine analysis. Photometers for microtiter plates are filter photometers.

Infrared Light Transmission Photometry

Spectrophotometry in infrared light is mainly used to study structure of substances, as given groups give absorption at defined wavelengths. Measurement in aqueous solution is generally not possible, as water absorbs infrared light strongly in some wavelength ranges. Therefore, infrared spectroscopy is either performed in the gaseous phase (for volatile substances) or with the substances pressed into tablets together with salts that are transparent in the infrared range. Potassium bromide (KBr) is commonly used for this purpose. The substance being tested is thoroughly mixed with specially purified KBr and pressed into a transparent tablet, that is placed in the beam of light. The analysis of the wavelength dependence is generally not done using a monochromator as it is in UV-Vis, but with the use of an interferometer. The interference pattern can be analyzed using a Fourier transform algorithm. In this way, the whole wavelength range can be analyzed simultaneously, saving time, and an interferometer is also less expensive than a monochromator. The light absorbed in the infrared region does not correspond to electronic excitation of the substance studied, but rather to different kinds of vibrational excitation. The vibrational excitations are characteristic of different groups in a molecule, that can in this way be identified. The infrared spectrum typically has very narrow absorption lines, which makes them unsuited for quantitative analysis but gives very detailed information about the molecules. The frequencies of the different modes of vibration varies with isotope, and therefore different isotopes give different peaks. This makes it possible also to study the isotopic composition of a sample with infrared spectrophotometry.

Atomic Absorption Photometry

Atomic absorption photometers are photometers that measure the light from a very hot flame. The solution to be analyzed is injected into the flame at a constant, known rate. Metals in the solution are present in atomic form in the flame. The monochromatic light in this type of photometer is generated by a discharge lamp where the discharge takes place in a gas with the metal to be determined. The discharge then emits light with wavelengths corresponding to the spectral lines of the metal. A filter may be used to isolate one of the main spectral lines of the metal to be analyzed. The light is absorbed by the metal in the flame, and the absorption is used to determine the concentration of the metal in the original solution.

POLARIMETER

A polarimeter is a scientific instrument used to measure the angle of rotation caused by passing polarized light through an optically active substance.

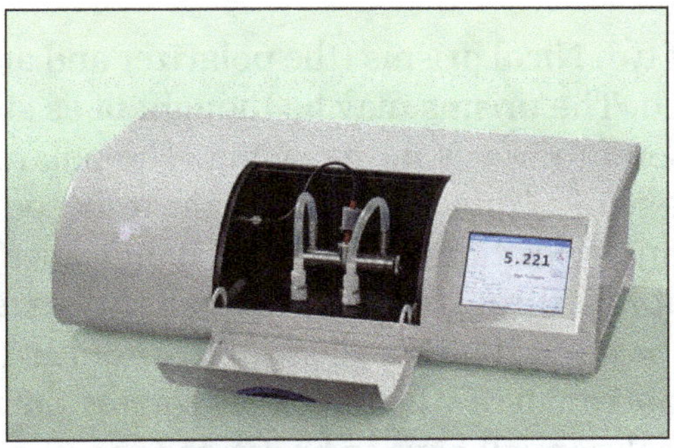

An automatic digital polarimeter.

Some chemical substances are optically active, and polarized (uni-directional) light will rotate either to the left (counter-clockwise) or right (clockwise) when passed through these substances. The amount by which the light is rotated is known as the angle of rotation. The angle of rotation is basically known as observed angle.

Measuring Principle

The ratio, the purity, and the concentration of two enantiomers can be measured via polarimetry. Enantiomers are characterized by their property to rotate the plane of linear polarized light. Therefore, those compounds are called optically active and their property is referred to as optical rotation. Light sources such as a light bulb, a light-emitting diode (LED), or the sun emit electromagnetic waves at the frequency of visible light. Their electric field oscillates in all possible planes relative to their direction of propagation. In contrast to that, the waves of linear-polarized light oscillate in parallel planes.

If light encounters a polarizer, only the part of the light that oscillates in the defined plane of the polarizer may pass through. That plane is called the plane of polarization. The plane of polarization is turned by optically active compounds. According to the direction in which the light is rotated, the enantiomer is referred to as dextro-rotatory or levo-rotatory.

The optical activity of enantiomers is additive. If different enantiomers exist together in one solution, their optical activity adds up. That is why racemates are optically inactive, as they nullify their clockwise and counter clockwise optical activities. The optical rotation is proportional to the concentration of the optically active substances in solution. Polarimeters may therefore be applied for concentration measurements of enantiomer-pure samples. With a known concentration of a sample, polarimeters may also be applied to determine the specific rotation when characterizing a new substance. The specific rotation is a physical property and defined as the optical rotation α at a path length l of 1 dm, a concentration c of 1g/100 mL, a temperature T (usually 20 °C) and a light wavelength λ (usually sodium D line at 589.3 nm):

$$[\alpha]_\lambda^T = \frac{100 \times \alpha}{l \times c},$$

This tells us how much the plane of polarization is rotated when the ray of light passes through a specific amount of optically active molecules of a sample. Therefore, the optical rotation depends on temperature, concentration, wavelength, path length, and the substance being analyzed.

Construction

The polarimeter is made up of two Nicol prisms (the polarizer and analyzer). The polarizer is fixed and the analyzer can be rotated. The prisms may be thought of as slits S1 and S2. The light waves may be considered to correspond to waves in the string. The polarizer S1 allows only those light waves which move in a single plane. This causes the light to become plane polarized. When the analyzer is also placed in a similar position it allows the light waves coming from the polarizer to pass through it. When it is rotated through the right angle no waves can pass through the right angle and the field appears to be dark. If now a glass tube containing an optically active solution is placed between the polarizer and analyzer the light now rotates through the plane of polarization through a certain angle, the analyzer will have to be rotated in same angle.

Operation

Polarimeters measure this by passing monochromatic light through the first of two polarising plates, creating a polarized beam. This first plate is known as the polarizer. This beam is then rotated as it passes through the sample. After passing through the sample, a second polarizer, known as the analyzer, rotates either via manual rotation or automatic detection of the angle. When the analyzer is rotated such that all the light or no light can pass through, then one can find the angle of rotation which is equal to the angle by which the analyser was rotated (θ) in the former case or (90-θ) in the latter case.

Types of Polarimeter

Laurent's Half-shade Polarimeter

When plane-polarised light passes through some crystals, the velocity of left-polarised light is different from that of the right-polarised light thus the crystals are said to have two refractive indices i.e. double refracting Construction: It consists of a monochromatic source S which is placed at focal point of a convex lens L. Just after the convex lens there is a Nicol Prism P which acts as a polariser. H is a half shade device which divides the field of polarised light emerging out of the Nicol P into two halves generally of unequal brightness. T is a glass tube in which optically active solution is filled. The light after passing through T is allowed to fall on the analyzing Nicol A which can be rotated about the axis of the tube. The rotation of analyser can be measured with the help of a scale C. Laurent's half-shade polarimeter is shown in figure.

Working: In order to understand the need of a half-shade device, let us suppose that half-shade device is not present. The position of the analyzer is so adjusted that the field of view is dark when tube is empty. The position of the analyzer is noted on circular scale. Now the tube is filled with optically active solution and it is set in its proper position. The optically active solution rotates the plane of polarization of the light emerging out of the polariser P by some angle. So the light is transmitted by analyzer A and the field of view of telescope becomes bright. Now the analyzer is rotated by a finite angle so that the field of view of telescope again become dark. This will happen only when the analyzer is rotated by the same angle by which plane of polarization of light is rotated by optically active solution.

The position of analyzer is again noted. The difference of the two readings will give you angle of rotation of plane of polarization.

A difficulty is faced in the above procedure that when analyzer is rotated for the total darkness, then it is attained gradually and hence it is difficult to find the exact position correctly for which complete darkness is obtained.

To overcome above difficulty half-shade device is introduced between polariser P and glass tube T.

Half Shade Device It consist of two semicircular plates ACB and ADB. One half ACB is made of glass while other half is made of quartz.

Both the halves are cemented together. The quartz is cut parallel to the optic axis. Thickness of the quartz is selected in such a way that it introduces a path difference of A/2 between ordinary and extraordinary ray. The thickness of the glass is selected in such a way that it absorbs the same amount of light as is absorbed by quartz half.

Consider that the vibration of polarisation is along OP. On passing through the glass half the vibrations remain along OP. But on passing through quartz half these vibrations will split into o- and £-components. The £-components are parallel to the optic axis while O- component is perpendicular to optic axis. The O-component travels faster in quartz and hence an emergence o-component will be along OD instead of along OC. Thus components OA and OD will combine to form a resultant vibration along OQ which makes same angle with optic axis as OP. Now if the Principal plane of the analyzing Nicol is parallel to OP then the light will pass through glass half unobstructed. Hence glass half will be brighter than quartz half or we can say that glass half will be bright and the quartz half will be dark. Similarly if principal plane of analyzing Nicol is parallel to OQ then quartz half will be bright and glass half will be dark.

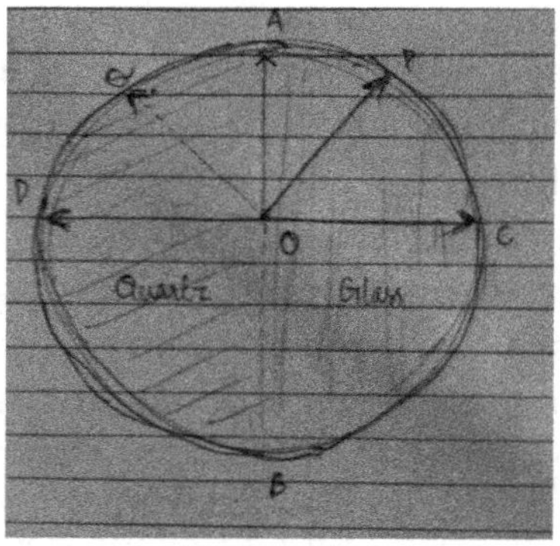

When the principal plane of analyzer is along AOB then both halves will be equally bright. On the other hand, if the principal plane of analyzer is along DOC. then both the halves will be equally dark.

Thus it is clear that if the analyzing Nicol is slightly disturbed from DOC then one half becomes brighter than the other. Hence by using half shade device, one can measure angle of rotation more accurately.

Determination of Specific Rotation

In order to determine specific rotation of an optically active substance (say sugar) the polarimeter tube T is first filled with pure water and analyzer is adjusted for equal darkness (Both the halves should be equally dark) point. The position of the analyzer is noted with the help of scale. Now the polarimeter tube is filled with sugar solution of known concentration and again the analyser is adjusted in such a way that again equally dark point is achieved. The position of the analyzer is again noted. The difference of the two readings will give you angle of rotation θ. Hence specific rotation S is determined by using the relation:

$$[S]t\ \lambda = \theta\ / LC,$$

The above procedure may be repeated for different concentration.

Biquartz Polarimeter

In biquartz polarimeters, a biquartz plate is used. Biquartz plate consists of two semi circular plates of quartz each of thickness 3.75 mm. One half consists of right-handed optically active quartz, while the other is left-handed optically active quartz.

Manual

The earliest polarimeters, which date back to the 1830s, required the user to physically rotate one polarizing element (the analyzer) whilst viewing through another static element (the detector). The detector was positioned at the opposite end of a tube containing the optically active sample, and the user used his/her eye to judge the "alignment" when least light was observed. The angle of rotation was then read from a simple fixed to the moving polariser to within a degree or so.

Although most manual polarimeters produced today still adopt this basic principle, the many developments applied to the original opto-mechanical design over the years have significantly improved measurement performance. The introduction of a half-wave plate increased "distinction sensitivity", whilst a precision glass scale with vernier drum facilitated the final reading to within ca. ±0.05°. Most modern manual polarimeters also incorporate a long-life yellow LED in place of the more costly sodium arc lamp as a light source.

Semi-automatic

Today, semi-automatic polarimeters are available. The operator views the image via a digital display adjusts the analyzer angle with electronic controls.

Fully Automatic

Fully automatic polarimeters are now available and simply require the user to press a button and wait for a digital readout. Fast automatic digital polarimeters yield an accurate result within a second, regardless of the rotation angle of the sample. In addition, they provide continuous measurement, facilitating High-performance liquid chromatography and other kinetic investigations.

Another feature of modern polarimeters is the Faraday modulator. The Faraday modulator creates an alternating current magnetic field. It oscillates the plane of polarization to enhance the detection accuracy by allowing the point of maximal darkness to be passed through again and again and thus be determined with even more accuracy.

As the temperature of the sample has a significant influence on the optical rotation of the sample, modern polarimeters have already included Peltier Elements to actively control the temperature. Special techniques like a temperature controlled sample tube reduce measuring errors and ease operation. Results can directly be transferred to computers or networks for automatic processing. Traditionally, accurate filling of the sample cell had to be checked outside the instrument, as an appropriate control from within the device was not possible. Nowadays a camera system allows accurate monitoring of the sample and filling conditions in the sample cell from inside the instrument. A telecentric camera gives a sharp image over the complete length of any sample cell placed within modern instruments. The online monitoring of the filling process ensures that no bubbles or particles obstruct the measurement. A picture can be saved together with the recorded data. Any temperature gradients, inhomogeneous sample distributions or air bubbles can immediately be recognized before measurement, so that potential errors caused by bubbles or particles are no longer an issue.

Sources of Error

The angle of rotation of an optically active substance can be affected by:

- Concentration of the sample.
- Wavelength of light passing through the sample (generally, angle of rotation and wavelength tend to be inversely proportional).
- Temperature of the sample (generally the two are directly proportional).
- Length of the sample cell (input by the user into most automatic polarimeters to ensure better accuracy).
- Filling conditions (bubbles, temperature and concentration gradients).

Most modern polarimeters have methods for compensating and controlling these errors.

Calibration

Polarimeters can be calibrated – or at least verified – by measuring a quartz plate, which is constructed to always read at a certain angle of optical rotation (usually +34°, but +17° and +8.5° are also popular, depending on the sample). Quartz plates are preferred by many users because solid samples are much less affected by variations in temperature, and do not need to be mixed on-demand like sucrose solutions.

Applications

Because many optically active chemicals such as tartaric acid, are stereoisomers, a polarimeter can be used to identify which isomer is present in a sample – if it rotates polarized light to the left, it is a levo-isomer, and to the right, a dextro-isomer. It can also be used to measure the ratio of enantiomers in solutions.

The optical rotation is proportional to the concentration of the optically active substances in solution. Polarimetry may therefore be applied for concentration measurements of enantiomer-pure samples. With a known concentration of a sample, polarimetry may also be applied to determine the specific rotation (a physical property) when characterizing a new substance.

Chemical Industry

Many chemicals exhibit a specific rotation as a unique property (an intensive property like refractive index or Specific gravity) which can be used to distinguish it. Polarimeters can identify unknown samples based on this if other variables such as concentration and length of sample cell length are controlled or at least known. This is used in the chemical industry.

By the same token, if the specific rotation of a sample is already known, then the concentration and purity of a solution containing it can be calculated.

Most automatic polarimeters make this calculation automatically, given input on variables from the user.

Food, Beverage and Pharmaceutical Industries

Concentration and purity measurements are especially important to determine product or ingredient quality in the food & beverage and pharmaceutical industries. Samples that display specific rotations that can be calculated for purity with a polarimeter include:

- Steroids
- Diuretics
- Antibiotics
- Narcotics
- Vitamins
- Analgesics
- Amino acids
- Essential oils
- Polymers
- Starches
- Sugars

Polarimeters are used in the sugar industry for determining quality of both juice from sugar cane and the refined sucrose. Often, the sugar refineries use a modified polarimeter with a flow cell (and used in conjunction with a refractometer) called a saccharimeter. These instruments use the International Sugar Scale, as defined by the International Commission for Uniform Methods of Sugar Analysis (ICUMSA).

REFRACTOMETER

A refractometer is a laboratory or field device for the measurement of an index of refraction (refractometry). The index of refraction is calculated from Snell's law while for mixtures, the index of refraction can be calculated from the composition of the material using several mixing rules such as the Gladstone–Dale relation and Lorentz–Lorenz equation.

Refractometry

Standard refractometers measure the extent of light refraction (as part of a refractive index) of transparent substances in either a liquid or solid state; this is then used in order to identify a liquid sample, analyse the sample's purity and determine the amount or concentration of dissolved substances within the sample. As light passes through the liquid from the air it will slow down and create a 'bending' illusion, the severity of the 'bend' will depend on the amount of substance dissolved in the liquid. For example, the amount of sugar in a glass of water.

Types of Refractometers

There are four main types of refractometers: traditional handheld refractometers, digital handheld refractometers, laboratory or Abbe refractometers (named for the instrument's inventor and based on Ernst Abbe's original design of the 'critical angle') and inline process refractometers. There is also the Rayleigh Refractometer used (typically) for measuring the refractive indices of gases.

In laboratory medicine, a refractometer is used to measure the total plasma protein in a blood sample and urine specific gravity in a urine sample.

In drug diagnostics, a refractometer is used to measure the specific gravity of human urine.

In gemology, the gemstone refractometer is one the fundamental pieces of equipment used in a gemological laboratory. Gemstones are transparent minerals and can therefore be examined using optical methods. Refractive index is a material constant, dependent on the chemical composition of a substance. The refractometer is used to help identify gem materials by measuring their refractive index, one of the principal properties used in determining the type of a gemstone. Due to the dependence of the refractive index on the wavelength of the light used (*i.e.* dispersion), the measurement is normally taken at the wavelength of the sodium line D-line (Na_D) of ~589 nm. This is either filtered out from daylight or generated with a monochromatic light-emitting diode (LED). Certain stones such as rubies, sapphires, tourmalines and topaz are optically anisotropic. They demonstrate birefringence based on the polarisation plane of the light. The two different refractive indexes are classified using a polarisation filter. Gemstone refractometers are available both as classic optical instruments and as electronic measurement devices with a digital display.

A wine grape grower with refractometer.

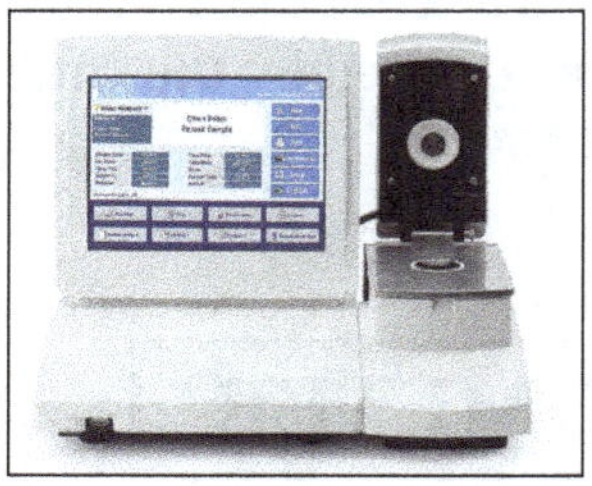

Automatic benchtop refractometer.

In marine aquarium keeping, a refractometer is used to measure the salinity and specific gravity of the water. In the automobile industry, a refractometer is used to measure the coolant concentration. In the machine industry, a refractometer is used to measure the amount of coolant concentrate that has been added to the water-based coolant for the machining process. In homebrewing, a brewing refractometer is used to measure the specific gravity before fermentation to determine the amount of fermentable sugars which will potentially be converted to alcohol.

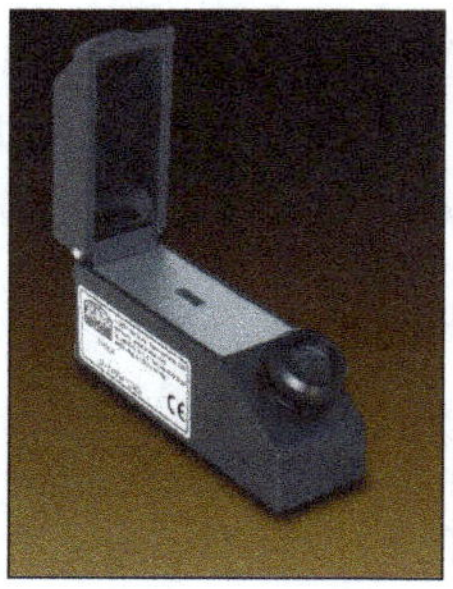

Gemology refractometer ER604 used to test light bending in gemstones.

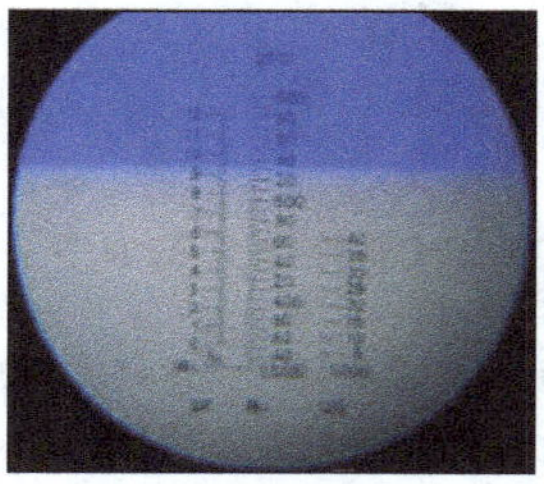

Density evaluation of abdominal fluid of a cat with feline infectious peritonitis by a refractometer.

Brix refractometers are often used by hobbyists for making preserves including jams, marmalades and honey. In beekeeping, a brix refractometer is used to measure the amount of water in honey.

Automatic Refractometers

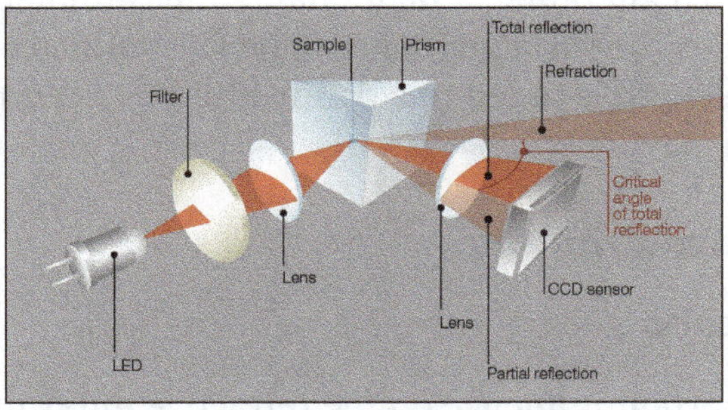

Schematic setup of an automatic refractometer.

In figure an LED light source is imaged under a wide range of angles onto a prism surface which is in contact with a sample. Depending on the difference in the refractive index between prism material and sample the light is partly transmitted or totally reflected. The critical angle of total reflection is determined by measuring the reflected light intensity as a function of the incident angle.

Automatic refractometers automatically measure the refractive index of a sample. The automatic measurement of the refractive index of the sample is based on the determination of the critical angle of total reflection. A light source, usually a long-life LED, is focused onto a prism surface via a lens system. An interference filter guarantees the specified wavelength. Due to focusing light to a spot at the prism surface, a wide range of different angles is covered. As shown in the figure *"Schematic setup of an automatic refractometer"* the measured sample is in direct contact with the measuring prism. Depending on its refractive index, the incoming light below the critical angle of total reflection is partly transmitted into the sample, whereas for higher angles of incidence the light is totally reflected. This dependence of the reflected light intensity from the incident angle is measured with a high-resolution sensor array. From the video signal taken with the CCD sensor the refractive index of the sample can be calculated. This method of detecting the angle of total reflection is independent on the sample properties. It is even possible to measure the refractive index of optically dense strongly absorbing samples or samples containing air bubbles or solid particles. Furthermore, only a few microliters are required and the sample can be recovered. This determination of the refraction angle is independent of vibrations and other environmental disturbances.

Influence of Wavelength

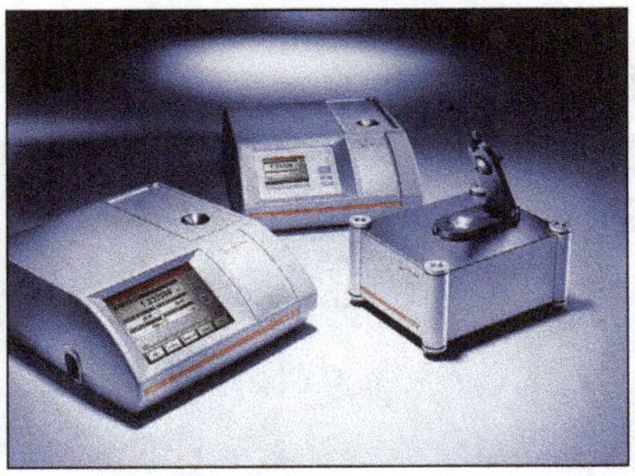

The refractive index of a given sample varies with wavelength for all materials. This dispersion relation is nonlinear and is characteristic for every material. In the visible range, a decrease of the refractive index comes with increasing wavelength. In glass prisms very little absorption is observable. In the infrared wavelength range several absorption maxima and fluctuations in the refractive index appear. To guarantee a high quality measurement with an accuracy of up to 0.00002 in the refractive index the wavelength has to be determined correctly. Therefore, in modern refractometers the wavelength is tuned to a bandwidth of +/-0.2 nm to ensure correct results for samples with different dispersions.

Modern Automatic Refractometers

Influence of Temperature

Temperature has a very important influence on the refractive index measurement. Therefore, the temperature of the prism and the temperature of the sample have to be controlled with high precision. There are several subtly-different designs for controlling the temperature; but there are some key factors common to all, such as high-precision temperature sensors and Peltier devices to control the temperature of the sample and the prism. The temperature control of these devices should be designed so that the variation in sample temperature is small enough that it will not cause a detectable refractive-index change.

External water baths were used in the past but are no longer needed.

Extended Possibilities of Automatic Refractometers

Automatic refractometers are microprocessor-controlled electronic devices. This means they can have a high degree of automation and also be combined with other measuring devices.

Flow Cells

There are different types of sample cells available, ranging from a flow cell for a few microliters to sample cells with a filling funnel for fast sample exchange without cleaning the measuring prism in between. The sample cells can also be used for the measurement of poisonous and toxic samples with minimum exposure to the sample. Micro cells require only a few microliters volume, assure good recovery of expensive samples and prevent evaporation of volatile samples or solvents. They can also be used in automated systems for automatic filling of the sample onto the refractometer prism. For convenient filling of the sample through a funnel, flow cells with a filling funnel are available. These are used for fast sample exchange in quality control applications.

Automatic Sample Feeding

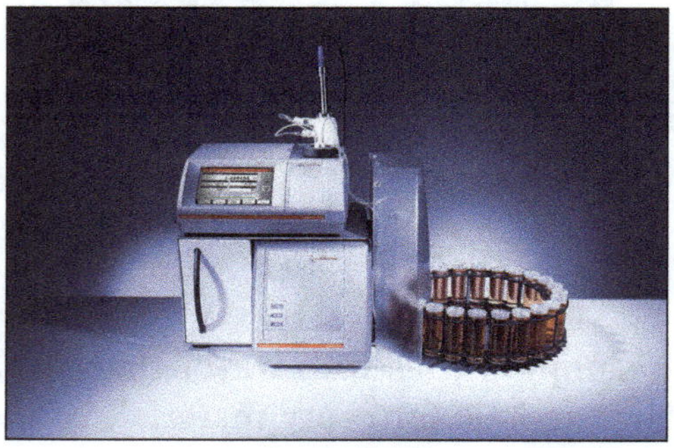

Automatic refractometer with sample changer for automatic measurement of a large number of samples.

Once an automatic refractometer is equipped with a flow cell, the sample can either be filled by means of a syringe or by using a peristaltic pump. Modern refractometers have the option of a built-in peristaltic pump. This is controlled via the instrument's software menu. A peristaltic pump opens the way to monitor batch processes in the laboratory or perform multiple measurements on one sample without any user interaction. This eliminates human error and assures a high sample throughput.

If an automated measurement of a large number of samples is required, modern automatic refractometers can be combined with an automatic sample changer. The sample changer is controlled by the refractometer and assures fully automated measurements of the samples placed in the vials of the sample changer for measurements.

Multiparameter Measurements

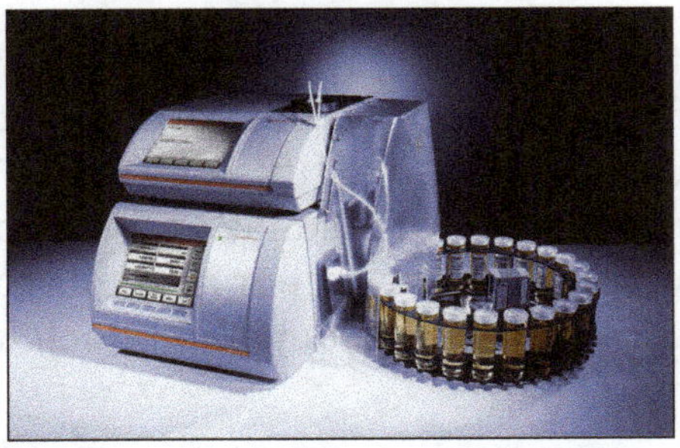

Measuring combination of an automatic refractometer and a density meter as widely used in the flavors and fragrances industry.

Today's laboratories do not only want to measure the refractive index of samples, but several additional parameters like density or viscosity to perform efficient quality control. Due to the microprocessor control and a number of interfaces, automatic refractometers are able to communicate with computers or other measuring devices, e.g. density meters, pH meters or viscosity meters, to store refractive index data and density data (and other parameters) into one database.

Software Features

Automatic refractometers do not only measure the refractive index, but offer a lot of additional software features, like:

- Instrument settings and configuration via software menu.
- Automatic data recording into a database.
- User-configurable data output.
- Export of measuring data into Microsoft Excel data sheets.
- Statistical functions.
- Predefined methods for different kinds of applications.
- Automatic checks and adjustments.
- Check if sufficient amount of sample is on the prism.
- Data recording only if the results are plausible.

Pharma Documentation and Validation

Refractometers are often used in pharmaceutical applications for quality control of raw intermediate and final products. The manufacturers of pharmaceuticals have to follow several international regulations like FDA 21 CFR Part 11, GMP, Gamp 5, which require a lot of documentation work. The manufacturers of automatic refractometers support these users providing instrument software fulfills the requirements of 21 CFR Part 11, with user levels, electronic signature and audit trail.

OPTICAL SPECTROMETER

An optical spectrometer (spectrophotometer, spectrograph or spectroscope) is an instrument used to measure properties of light over a specific portion of the electromagnetic spectrum, typically used in spectroscopic analysis to identify materials. The variable measured is most often the light's intensity but could also, for instance, be the polarization state. The independent variable is usually the wavelength of the light or a unit directly proportional to the photon energy, such as reciprocal centimeters or electron volts, which has a reciprocal relationship to wavelength.

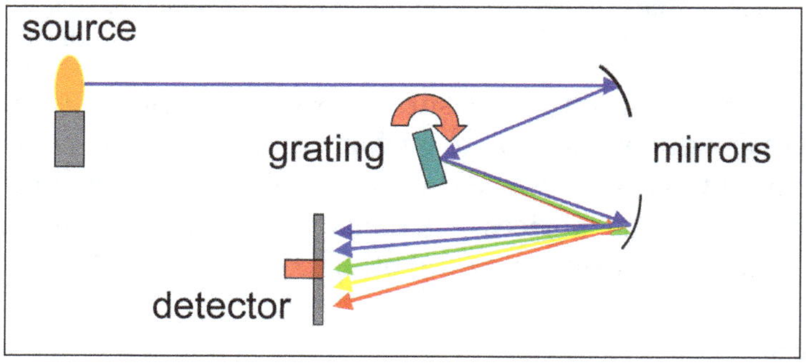

Grating spectrometer schematic.

A spectrometer is used in spectroscopy for producing spectral lines and measuring their wavelengths and intensities. Spectrometers may also operate over a wide range of non-optical wavelengths, from gamma rays and X-rays into the far infrared. If the instrument is designed to measure the spectrum in absolute units rather than relative units, then it is typically called a spectrophotometer. The majority of spectrophotometers are used in spectral regions near the visible spectrum.

In general, any particular instrument will operate over a small portion of this total range because of the different techniques used to measure different portions of the spectrum. Below optical frequencies (that is, at microwave and radio frequencies), the spectrum analyzer is a closely related electronic device.

Spectrometers are used in many fields. For example, they are used in astronomy to analyze the radiation from astronomical objects and deduce chemical composition. The spectrometer uses a prism or a grating to spread the light from a distant object into a spectrum. This allows astronomers to detect many of the chemical elements by their characteristic spectral fingerprints. If the object is glowing by itself, it will show spectral lines caused by the glowing gas itself. These lines are named for the elements which cause them, such as the hydrogen alpha, beta, and gamma lines. Chemical compounds may also be identified by absorption. Typically these are dark bands in specific locations in the spectrum caused by energy being absorbed as light from other objects passes through a gas cloud. Much of our knowledge of the chemical makeup of the universe comes from spectra.

Spectroscopes

Spectroscopes are often used in astronomy and some branches of chemistry. Early spectroscopes were simply prisms with graduations marking wavelengths of light. Modern spectroscopes generally use a diffraction grating, a movable slit, and some kind of photodetector, all automated and controlled by a computer.

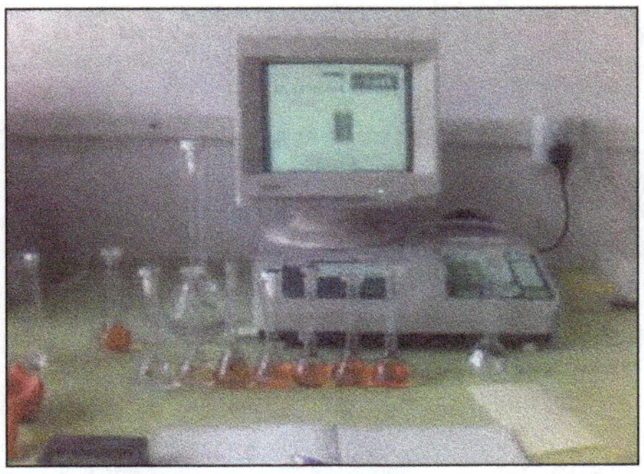

Spectroscope.

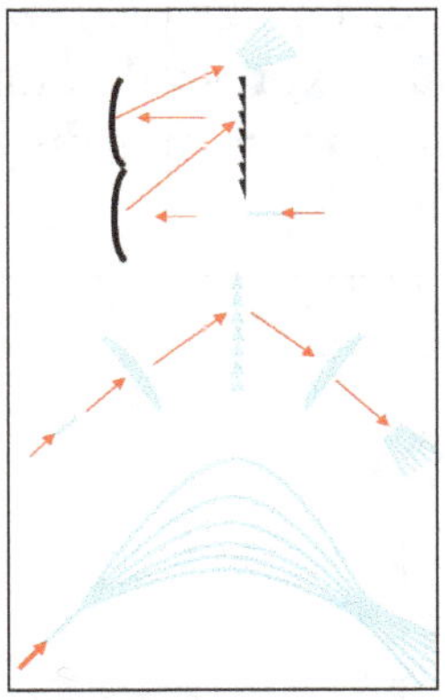

Comparison of different diffraction based spectrometers: Reflection optics, refraction optics, fiber/integrated optics.

Joseph von Fraunhofer developed the first modern spectroscope by combining a prism, diffraction slit and telescope in a manner that increased the spectral resolution and was reproducible in other laboratories. Fraunhofer also went on to invent the first diffraction spectroscope. Gustav Robert Kirchhoff and Robert Bunsen discovered the application of spectroscopes to chemical analysis and used this approach to discover caesium and rubidium. Kirchhoff and Bunsen's analysis also enabled a chemical explanation of stellar spectra, including Fraunhofer lines.

When a material is heated to incandescence it emits light that is characteristic of the atomic makeup of the material. Particular light frequencies give rise to sharply defined bands on the scale which can be thought of as fingerprints. For example, the element sodium has a very characteristic double yellow band known as the Sodium D-lines at 588.9950 and 589.5924 nanometers, the color of which will be familiar to anyone who has seen a low pressure sodium vapor lamp.

In the original spectroscope design in the early 19th century, light entered a slit and a collimating lens transformed the light into a thin beam of parallel rays. The light then passed through a prism (in hand-held spectroscopes, usually an Amici prism) that refracted the beam into a spectrum because different wavelengths were refracted different amounts due to dispersion. This image was then viewed through a tube with a scale that was transposed upon the spectral image, enabling its direct measurement.

With the development of photographic film, the more accurate spectrograph was created. It was based on the same principle as the spectroscope, but it had a camera in place of the viewing tube. In recent years, the electronic circuits built around the photomultiplier tube have replaced the camera, allowing real-time spectrographic analysis with far greater accuracy. Arrays of photosensors are also used in place of film in spectrographic systems. Such spectral analysis, or spectroscopy, has become an important scientific tool for analyzing the composition of unknown material and for studying astronomical phenomena and testing astronomical theories.

In modern spectrographs in the UV, visible, and near-IR spectral ranges, the spectrum is generally given in the form of photon number per unit wavelength (nm or μm), wavenumber (μm^{-1}, cm^{-1}), frequency (THz), or energy (eV), with the units indicated by the abscissa. In the mid- to far-IR, spectra are typically expressed in units of Watts per unit wavelength (μm) or wavenumber (cm^{-1}). In many cases, the spectrum is displayed with the units left implied (such as "digital counts" per spectral channel).

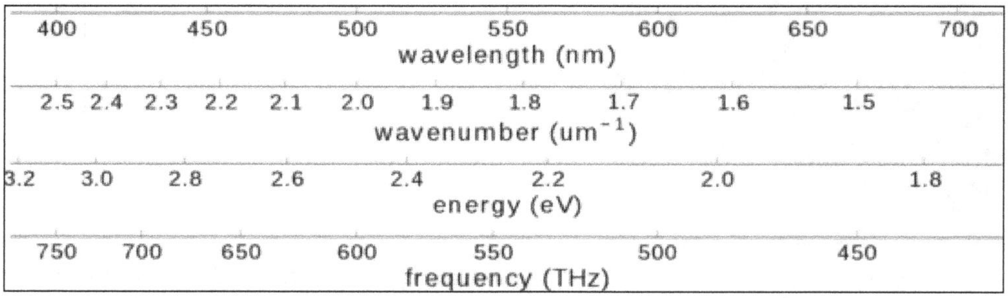

A comparison of the four abscissa types typically used for visible spectrometers.

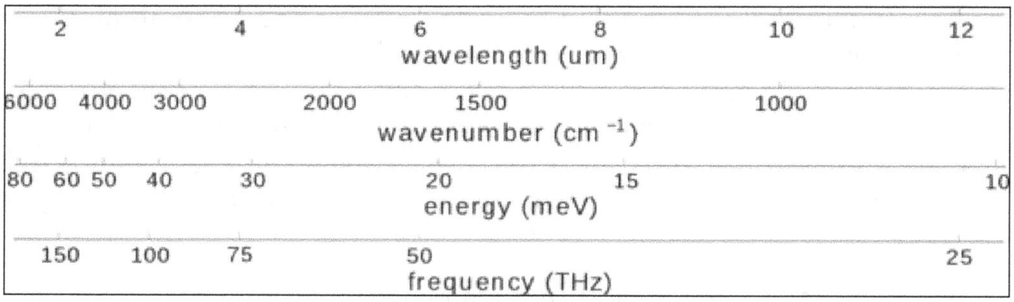

A comparison of the four abscissa types typically used for infrared spectrometers.

Spectrographs

A spectrograph is an instrument that separates light by its wavelengths and records this data. A spectrograph typically has a multi-channel detector system or camera that detects and records the spectrum of light.

The term was first used in 1876 by Dr. Henry Draper when he invented the earliest version of this device, and which he used to take several photographs of the spectrum of Vega. This earliest version of the spectrograph was cumbersome to use and difficult to manage.

A very simple spectroscope based on a prism.

Horizontal Solar Spectrograph at the Czech Astronomical Institute in Ondřejov, Czech Republic.

There are several kinds of machines referred to as *spectrographs*, depending on the precise nature of the waves. The first spectrographs used photographic paper as the detector. The star spectral classification and discovery of the main sequence, Hubble's law and the Hubble sequence were all made with spectrographs that used photographic paper. The plant pigment phytochrome was discovered using a spectrograph that used living plants as the detector. More recent spectrographs use electronic detectors, such as CCDs which can be used for both visible and UV light. The exact choice of detector depends on the wavelengths of light to be recorded.

A spectrograph is sometimes called polychromator, as an analogy to monochromator.

Stellar and Solar Spectrograph

The first spectrographs used photographic paper as the detector. The star spectral classification and discovery of the main sequence, Hubble's law and the Hubble sequence were all made with spectrographs that used photographic paper. The plant pigment phytochrome was discovered using a spectrograph that used living plants as the detector. More recent spectrographs use electronic detectors, such as CCDs which can be used for both visible and UV light. The exact choice of detector depends on the wavelengths of light to be recorded.

The forthcoming James Webb Space Telescope will contain both a near-infrared spectrograph (NIRSpec) and a mid-infrared spectrograph (MIRI).

Echelle Spectrograph

An Echelle spectrograph uses two diffraction gratings, rotated 90 degrees with respect to each other and placed close to one another. Therefore, an entrance point and not a slit is used and a 2d CCD-chip records the spectrum. Usually one would guess to retrieve a spectrum on the diagonal, but when both gratings have a wide spacing and one is blazed so that only the first order is visible and the other is blazed that a lot of higher orders are visible, one gets a very fine spectrum nicely folded onto a small common CCD-chip. The small chip also means that the collimating optics need not to be optimized for coma or astigmatism, but the spherical aberration can be set to zero.

OPTICAL MAPPING

Optical mapping is a technique for constructing ordered, genome-wide, high-resolution restriction maps from single, stained molecules of DNA, called "optical maps". By mapping the location of restriction enzyme sites along the unknown DNA of an organism, the spectrum of resulting DNA fragments collectively serves as a unique "fingerprint" or "barcode" for that sequence. Originally developed by Dr. David C. Schwartz and his lab at NYU in the 1990s this method has since been integral to the assembly process of many large-scale sequencing projects for both microbial and eukaryotic genomes. Later technologies use DNA melting, DNA competitive binding or enzymatic labelling in order to create the optical mappings.

Technology

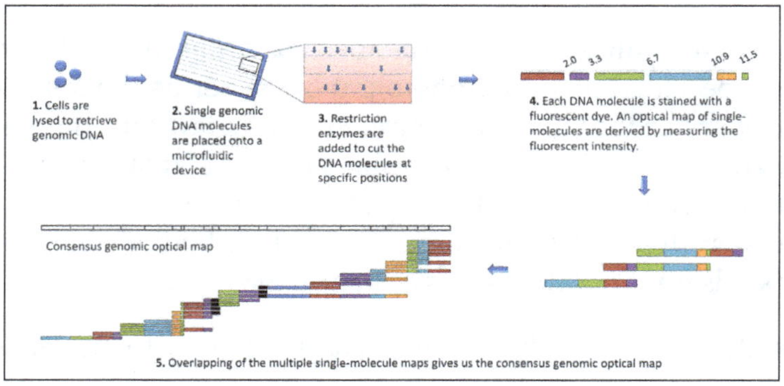

The optical mapping workflow.

The modern optical mapping platform works as follows:

- Genomic DNA is obtained from lysed cells, and randomly sheared to produce a "library" of large genomic molecules for optical mapping.
- A single molecule of DNA is stretched (or elongated) and held in place on a slide under a fluorescent microscope due to charge interactions.
- The DNA molecule is digested by added restriction enzymes, which cleave at specific digestion sites. The resulting molecule fragments remain attached to the surface. The fragment ends at the cleavage sites are drawn back (due to elasticity of linearized DNA), leaving gaps which are identifiable under the microscope.
- DNA fragments stained with intercalating dye are visualized by fluorescence microscopy and are sized by measuring the integrated fluorescence intensity. This produces an optical map of single molecules.
- Individual optical maps are combined to produce a consensus, genomic optical map.

Early System

DNA molecules were fixed on molten agarose developed between a cover slip and a microscope slide. Restriction enzyme was pre-mixed with the molten agarose before DNA placement and cleavage was triggered by addition of magnesium.

Using Charged Surfaces

Rather than being immobilized within a gel matrix, DNA molecules were held in place by electrostatic interactions on a positively charged surface. Resolution improved such that fragments from ~30kb to as small as 800bp could sized.

Automated System

This involved the development and integration of an automated spotting system to spot multiple single molecules on a slide (like a microarray) for parallel enzymatic processing, automated fluorescence microscopy for image acquisition, image procession vision to handle images, algorithms for optical map construction, cluster computing for processing large amounts of data.

High-throughput System using Microfluidics

Observing that microarrays spotted with single molecules did not work well for large genomic DNA molecules, microfluidic devices using soft lithography possessing a series of parallel microchannels were developed.

Next-generation System using Nanocoding Technology

An improvement on optical mapping, called "Nanocoding", has potential to boost throughput by trapping elongated DNA molecules in nanoconfinements.

Comparisons

Other Mapping Techniques

The advantage of OM over traditional mapping techniques is that it preserves the order of the DNA fragment, whereas the order needs to be reconstructed using restriction mapping. In addition, since maps are constructed directly from genomic DNA molecules, cloning or PCR artifacts are avoided. However, each OM process is still affected by false positive and negative sites because not all restriction sites are cleaved in each molecule and some sites may be incorrectly cut. In practice, multiple optical maps are created from molecules of the same genomic region, and an algorithm is used to determine the best consensus map.

Other Genome Analysis Methods

There are a variety of approaches to identifying large-scale genomic variations (such as indels, duplications, inversions, translocations) between genomes. Other categories of methods include using microarrays, pulsed-field gel electrophoresis, cytogenetics and paired-end tags.

Uses

Initially, the optical mapping system has been used to construct whole-genome restriction maps of bacteria, parasites, and fungi. It has also been used to scaffold and validate bacterial genomes. To serve as scaffolds for assembly, assembled sequence contigs can be scanned for restriction sites *in silico* using known sequence data and aligning them to the assembled genomic optical map.

Commercial company, Opgen has provided optical mappings for microbial genomes. For larger eukaryotic genomes, only the David C. Schwartz lab (now at Madison-Wisconsin) has produced optical maps for mouse, human, rice, and maize.

Optical Sequencing

Optical sequencing is a single molecule DNA sequencing technique that follows sequence-by-synthesis and uses optical mapping technology. Similar to other single molecular sequencing approaches such as SMRT sequencing, this technique analyzes a single DNA molecule, rather than amplify the initial sample and sequence multiple copies of the DNA. During synthesis, fluorochrome-labeled nucleotides are incorporated through the use of DNA polymerases and tracked by fluorescence microscopy. This technique was originally proposed by David C. Schwartz and Arvind Ramanathan in 2003.

Optical Sequencing Cycle

The following is an overview of each cycle in the optical sequencing process.

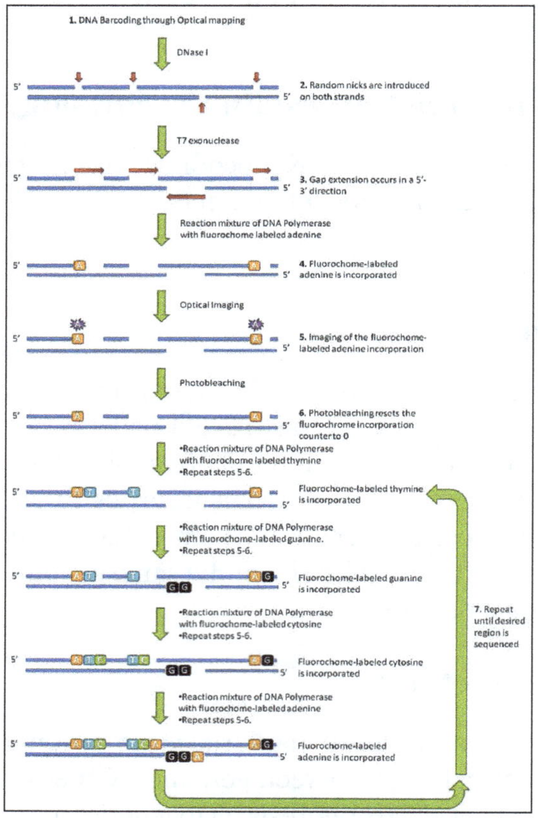

The optical sequencing cycle.

Step 1: DNA Barcoding

Cells are lysed to release genomic DNA. These DNA molecules are untangled, placed onto optical mapping surface containing microfluidic channels and the DNA is allowed to flow through the channels. These molecules are then barcoded by restriction enzymes to allow for genomic localization through the technique of optical mapping.

Step 2: Template Nicking

DNase I is added to randomly nick the mounted DNA molecules. A wash is then performed to remove the DNase I. The mean number of nicks that occur per template is dependent on the concentration of DNase I as well as the incubation time.

Step 3: Gap Formation

T7 exonuclease is added which uses the nicks in the DNA molecules to expand the gaps in a 5'–3' direction. Amount of T7 exonuclease must be carefully controlled to avoid overly high levels of double-stranded breaks.

Step 4: Fluorochrome Incorporation

DNA polymerase is used to incorporate fluorochrome-labelled nucleotides (FdNTPs) into the multiple gapped sites along each DNA molecule. During each cycle, the reaction mixture contains a single type of FdNTP and allows for multiple additions of that nucleotide type. Various washes are then performed to remove unincorporated fdNTPs in preparation for imaging and the next cycle of FdNTP addition.

Step 5: Imaging

This step counts the number of incorporated fluorochrome-labeled nucleotides at the gap regions using fluorescence microscopy.

Step 6: Photobleaching

The laser illumination that is used to excite the fluorochrome is also used here to destroy the fluorochrome signal. This essentially resets the fluorochrome counter, and prepares the counter for the next cycle. This step is a unique aspect of optical sequencing as it does not actually remove the fluorochrome label of the nucleotide after its incorporation. not removing the fluorochrome label makes sequencing more economical, but it results in the need to incorporate fluorochrome labels consecutively which can result in problems due to the bulkiness of the labels.

Step 7: Repeat Steps 4–6

Steps 4-6 are repeated with step 4 using a reaction mixture that contains a different fluorochrome-labeled nucleotide (FdNTP) each time. This is repeated until the desired region is sequenced.

Optimization Strategies

Selection of an appropriate DNA polymerase is critical to the efficiency of the base addition step and must meet several criteria:

- Ability to efficiently incorporate FdNTP at consecutive positions.
- Lack of 3'–5' exonuclease and proofreading activity to prevent the removal newly incorporated FdNTP.

- High fidelity to minimize mis-incorporations.
- Good activity on templates which are mounted to surfaces (e.g. optical mapping surface).

In addition, different polymerase preference for different fluorochromes, linker length on fluorochrome-nucleotides, and buffer compositions are also important factors to be considered to optimize the base addition process and maximize number of consecutive FdNTP incorporations.

Advantages

Single-molecule Analysis

Since minimal DNA sample required, time-consuming and costly amplification step is avoided to streamline sample preparation process.

Large DNA Molecule Templates vs. Short DNA Molecule Templates

While most next generation sequencing technologies aim of massive amounts of smalls sequence reads, these small sequence reads make de novo sequencing efforts and genome repeat regions difficult to comprehend. Optical sequencing uses large DNA molecule templates (~500 kb) for sequencing and these offer several advantages over small templates:

- These large DNA templates can be "DNA barcoded" to determine their genomic localization with confidence. Therefore, any sequence reads that are taken from the large template can be mapped onto the genome with a high degree of confidence. More importantly, sequence reads from high repeat regions can placed with a greater degree of confidence whereas the short reads suffer from mapping uncertainty in high repeat regions. Special algorithms and software such as optical mapping and nanocoding have been developed to align single-molecule barcodes with a reference genome.

- Multiple sequence reads from the same large template molecule. These multiple sequence reads reduce the complexity of de novo assembly, disambiguate genomic rearrangement regions, and "intrinsically free from any assembly errors".

- Molecular barcoding of large DNA molecular templates with sequence acquisition provides broad and specific genomic analyses.

Disadvantages

- Single molecule DNA sequencing requires a high level of precision to match the confidence from the redundant read coverage provided by current next-generation sequencing technologies.

- Nicks on both strands at similar positions resulting in low template during sequence-by-synthesis.

- Fluorochrome-labeled nucleotides are not removed after incorporation and because of these bulky labels, multiple incorporation might be difficult.

LASERS

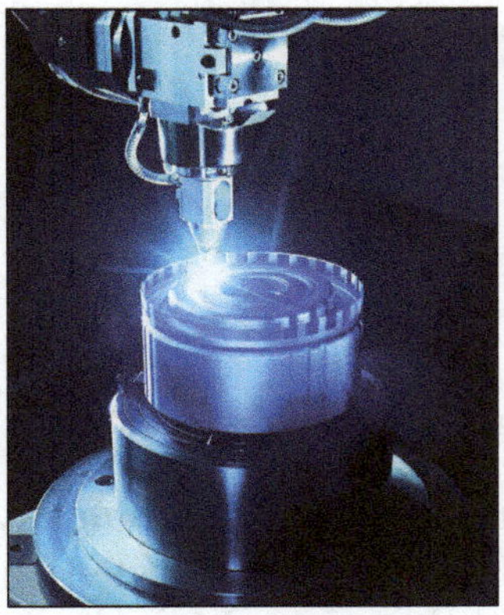

A laser beam used for welding.

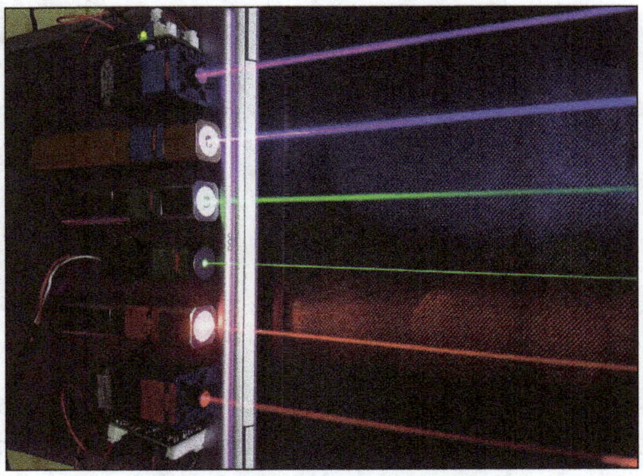

Red (660 & 635 nm), green (532 & 520 nm) and blue-violet (445 & 405 nm) lasers.

A laser is a device that emits light through a process of optical amplification based on the stimulated emission of electromagnetic radiation. The term "laser" originated as an acronym for "light amplification by stimulated emission of radiation". The first laser was built in 1960 by Theodore H. Maiman at Hughes Research Laboratories, based on theoretical work by Charles Hard Townes and Arthur Leonard Schawlow.

A laser differs from other sources of light in that it emits light *coherently*. Spatial coherence allows a laser to be focused to a tight spot, enabling applications such as laser cutting and lithography. Spatial coherence also allows a laser beam to stay narrow over great distances (collimation), enabling applications such as laser pointers and lidar. Lasers can also have high temporal coherence, which allows them to emit light with a very narrow spectrum, i.e., they can emit a single color of light. Alternatively, temporal coherence can be used to produce pulses of light with a broad spectrum but durations as short as a femtosecond ("ultrashort pulses").

Lasers are used in optical disk drives, laser printers, barcode scanners, DNA sequencing instruments, fiber-optic and free-space optical communication, laser surgery and skin treatments, cutting and welding materials, military and law enforcement devices for marking targets and measuring range and speed, and in laser lighting displays for entertainment. They have been used for car headlamps on luxury cars, by using a blue laser and a phosphor to produce highly directional white light.

Fundamentals

Lasers are distinguished from other light sources by their coherence. Spatial coherence is typically expressed through the output being a narrow beam, which is diffraction-limited. Laser beams can be focused to very tiny spots, achieving a very high irradiance, or they can have very low divergence in order to concentrate their power at a great distance. Temporal (or longitudinal) coherence implies a polarized wave at a single frequency, whose phase is correlated over a relatively great distance (the coherence length) along the beam. A beam produced by a thermal or other incoherent light source has an instantaneous amplitude and phase that vary randomly with respect to time and position, thus having a short coherence length.

Modern telescopes use laser technologies to compensate for the blurring effect of the Earth's atmosphere.

Lasers are characterized according to their wavelength in a vacuum. Most "single wavelength" lasers actually produce radiation in several *modes* with slightly different wavelengths. Although temporal coherence implies monochromaticity, there are lasers that emit a broad spectrum of light or emit different wavelengths of light simultaneously. Some lasers are not single spatial mode and have light beams that diverge more than is required by the diffraction limit. All such devices are classified as "lasers" based on their method of producing light, i.e., stimulated emission. Lasers are employed where light of the required spatial or temporal coherence can not be produced using simpler technologies.

Design

A laser consists of a gain medium, a mechanism to energize it, and something to provide optical feedback. The gain medium is a material with properties that allow it to amplify light by way of stimulated emission. Light of a specific wavelength that passes through the gain medium is amplified (increases in power).

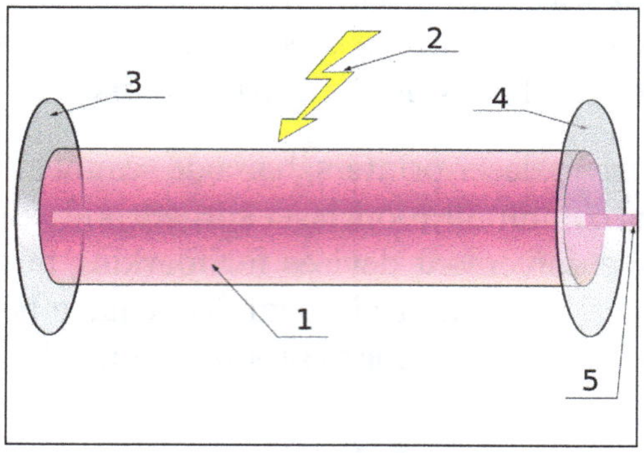

Components of a typical laser: 1. Gain medium, 2. Laser pumping energy, 3. High reflector, 4. Output coupler, 5. Laser beam.

For the gain medium to amplify light, it needs to be supplied with energy in a process called pumping. The energy is typically supplied as an electric current or as light at a different wavelength. Pump light may be provided by a flash lamp or by another laser.

The most common type of laser uses feedback from an optical cavity—a pair of mirrors on either end of the gain medium. Light bounces back and forth between the mirrors, passing through the gain medium and being amplified each time. Typically one of the two mirrors, the output coupler, is partially transparent. Some of the light escapes through this mirror. Depending on the design of the cavity (whether the mirrors are flat or curved), the light coming out of the laser may spread out or form a narrow beam. In analogy to electronic oscillators, this device is sometimes called a laser oscillator.

Most practical lasers contain additional elements that affect properties of the emitted light, such as the polarization, wavelength, and shape of the beam.

Laser Physics

Electrons and how they interact with electromagnetic fields are important in our understanding of chemistry and physics.

Stimulated Emission

In the classical view, the energy of an electron orbiting an atomic nucleus is larger for orbits further from the nucleus of an atom. However, quantum mechanical effects force electrons to take on discrete positions in orbitals. Thus, electrons are found in specific energy levels of an atom, two of which are shown below:

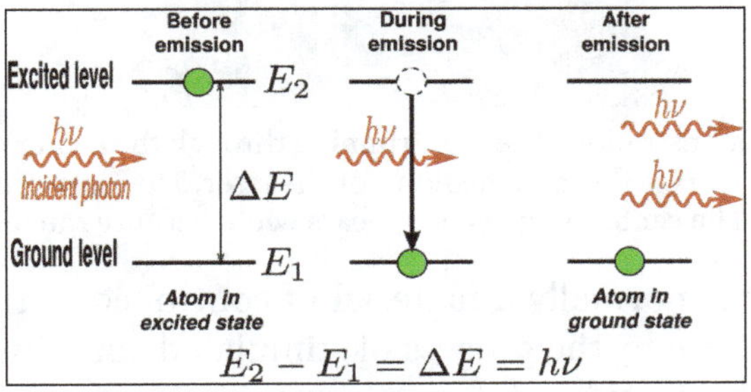

When an electron absorbs energy either from light (photons) or heat (phonons), it receives that incident quantum of energy. But transitions are only allowed in between discrete energy levels such as the two shown above. This leads to emission lines and absorption lines.

When an electron is excited from a lower to a higher energy level, it will not stay that way forever. An electron in an excited state may decay to a lower energy state which is not occupied, according to a particular time constant characterizing that transition. When such an electron decays without external influence, emitting a photon, that is called "spontaneous emission". The phase associated with the photon that is emitted is random. A material with many atoms in such an excited state may thus result in radiation which is very spectrally limited (centered around one wavelength of light), but the individual photons would have no common phase relationship and would emanate in random directions. This is the mechanism of fluorescence and thermal emission.

An external electromagnetic field at a frequency associated with a transition can affect the quantum mechanical state of the atom. As the electron in the atom makes a transition between two stationary states (neither of which shows a dipole field), it enters a transition state which does have a dipole field, and which acts like a small electric dipole, and this dipole oscillates at a characteristic frequency. In response to the external electric field at this frequency, the probability of the atom entering this transition state is greatly increased. Thus, the rate of transitions between two stationary states is enhanced beyond that due to spontaneous emission. Such a transition to the higher state is called absorption, and it destroys an incident photon (the photon's energy goes into powering the increased energy of the higher state). A transition from the higher to a lower energy state, however, produces an additional photon; this is the process of stimulated emission.

Gain Medium and Cavity

The gain medium is put into an excited state by an external source of energy. In most lasers this medium consists of a population of atoms which have been excited into such a state by means of an outside light source, or an electrical field which supplies energy for atoms to absorb and be transformed into their excited states.

A helium–neon laser demonstration. The glow running through the center of the tube is an electric discharge. This glowing plasma is the gain medium for the laser. The laser produces a tiny, intense spot on the screen to the right. The center of the spot appears white because the image is overexposed there.

The gain medium of a laser is normally a material of controlled purity, size, concentration, and shape, which amplifies the beam by the process of stimulated emission described above. This material can be of any state: gas, liquid, solid, or plasma. The gain medium absorbs pump energy, which raises some electrons into higher-energy ("excited") quantum states. Particles can interact with light by either absorbing or emitting photons. Emission can be spontaneous or stimulated. In the latter case, the photon is emitted in the same direction as the light that is passing by. When the number of particles in one excited state exceeds the number of particles in some lower-energy state, population inversion is achieved and the amount of stimulated emission due to light that passes through is larger than the amount of absorption. Hence, the light is amplified. By itself, this makes an optical amplifier. When an optical amplifier is placed inside a resonant optical cavity, one obtains a laser oscillator.

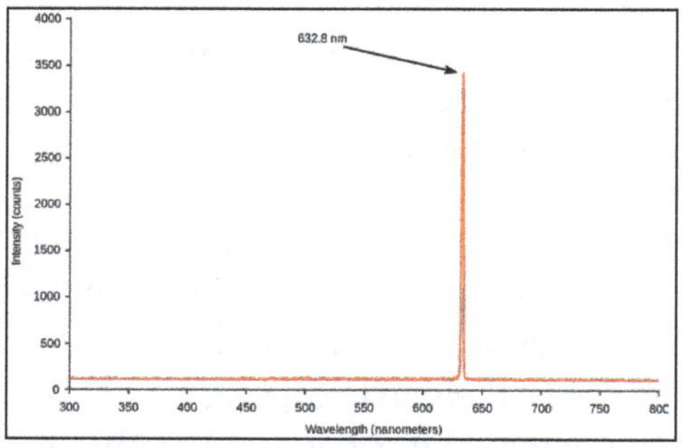

Spectrum of a helium–neon laser. The actual bandwidth is much narrower than shown; the spectrum is limited by the measuring apparatus.

In a few situations it is possible to obtain lasing with only a single pass of EM radiation through the gain medium, and this produces a laser beam without any need for a resonant or reflective cavity. Thus, reflection in a resonant cavity is usually required for a laser, but is not absolutely necessary.

The optical resonator is sometimes referred to as an "optical cavity", but this is a misnomer: lasers use open resonators as opposed to the literal cavity that would be employed at microwave frequencies in a maser. The resonator typically consists of two mirrors between which a coherent beam of light travels in both directions, reflecting back on itself so that an average photon will pass through the gain medium repeatedly before it is emitted from the output aperture or lost to diffraction or absorption. If the gain (amplification) in the medium is larger than the resonator losses, then the power of the recirculating light can rise exponentially. But each stimulated emission event returns an atom from its excited state to the ground state, reducing the gain of the medium. With increasing beam power the net gain (gain minus loss) reduces to unity and the gain medium is said to be saturated. In a continuous wave (CW) laser, the balance of pump power against gain saturation and cavity losses produces an equilibrium value of the laser power inside the cavity; this equilibrium determines the operating point of the laser. If the applied pump power is too small, the gain will never be sufficient to overcome the cavity losses, and laser light will not be produced. The minimum pump power needed to begin laser action is called the *lasing threshold*. The gain medium will amplify any photons passing through it, regardless of direction; but only the photons in a spatial mode supported by the resonator will pass more than once through the medium and receive substantial amplification.

The Light Emitted

In most lasers, lasing begins with stimulated emission amplifying random spontaneously emitted photons present in the gain medium. Stimulated emission produces light that matches the input signal in wavelength, phase, and polarization. This, combined with the filtering effect of the optical resonator gives laser light its characteristic coherence, and may give it uniform polarization and monochromaticity, depending on the resonator's design. Some lasers use a separate injection seeder to start the process off with a beam that is already highly coherent. This can produce beams with a narrower spectrum than would otherwise be possible.

Many lasers produce a beam that can be approximated as a Gaussian beam; such beams have the minimum divergence possible for a given beam diameter. Some lasers, particularly high-power ones, produce multimode beams, with the transverse modes often approximated using Hermite–Gaussian or Laguerre-Gaussian functions. Some high power lasers use a flat-topped profile known as a "tophat beam". Unstable laser resonators (not used in most lasers) produce fractal-shaped beams. Specialized optical systems can produce more complex beam geometries, such as Bessel beams and optical vortexes.

Near the "waist" (or focal region) of a laser beam, it is highly *collimated*: the wavefronts are planar, normal to the direction of propagation, with no beam divergence at that point. However, due to diffraction, that can only remain true well within the Rayleigh range. The beam of a single transverse mode (gaussian beam) laser eventually diverges at an angle which varies inversely with the beam diameter, as required by diffraction theory. Thus, the "pencil beam" directly generated by a common helium–neon laser would spread out to a size of perhaps 500 kilometers when shone on the Moon (from the distance of the earth). On the other hand, the light from a semiconductor laser typically exits the tiny crystal with a large divergence: up to 50°. However even such a divergent beam can be transformed into a similarly collimated beam by means of a lens system, as is always included, for instance, in a laser pointer whose light originates from a laser diode. That is possible due to the light being of a single spatial mode. This unique property of laser light, spatial coherence, cannot be replicated using standard light sources (except by discarding most of the light) as can be appreciated by comparing the beam from a flashlight (torch) or spotlight to that of almost any laser.

A laser beam profiler is used to measure the intensity profile, width, and divergence of laser beams.

Diffuse reflection of a laser beam from a matte surface produces a speckle pattern with interesting properties.

Quantum vs. Classical Emission Processes

The mechanism of producing radiation in a laser relies on stimulated emission, where energy is extracted from a transition in an atom or molecule. This is a quantum phenomenon discovered by Einstein who derived the relationship between the A coefficient describing spontaneous emission and the B coefficient which applies to absorption and stimulated emission. However, in the case of the free electron laser, atomic energy levels are not involved; it appears that the operation of this rather exotic device can be explained without reference to quantum mechanics.

Continuous and Pulsed Modes of Operation

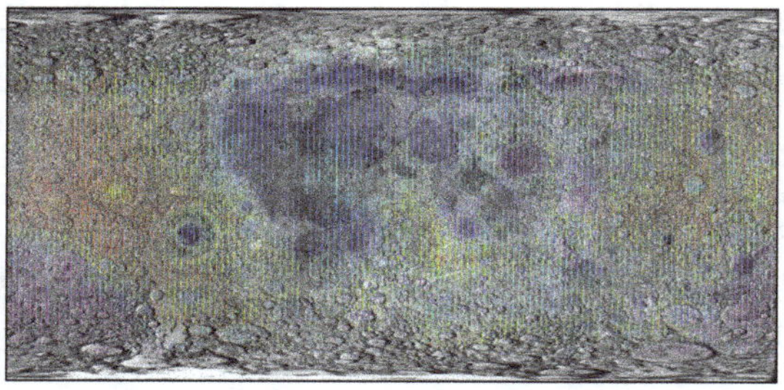

Lidar measurements of lunar topography made by Clementine mission.

Laserlink point to point optical wireless network.

Mercury Laser Altimeter (MLA) of the MESSENGER spacecraft.

A laser can be classified as operating in either continuous or pulsed mode, depending on whether the power output is essentially continuous over time or whether its output takes the form of pulses of light on one or another time scale. Of course even a laser whose output is normally continuous can be intentionally turned on and off at some rate in order to create pulses of light. When the modulation rate is on time scales much slower than the cavity lifetime and the time period over which energy can be stored in the lasing medium or pumping mechanism, then it is still classified as a "modulated" or "pulsed" continuous wave laser. Most laser diodes used in communication systems fall in that category.

Continuous Wave Operation

Some applications of lasers depend on a beam whose output power is constant over time. Such a laser is known as *continuous wave* (*CW*). Many types of lasers can be made to operate in continuous wave mode to satisfy such an application. Many of these lasers actually lase in several longitudinal modes at the same time, and beats between the slightly different optical frequencies of those oscillations will, in fact, produce amplitude variations on time scales shorter than the round-trip time (the reciprocal of the frequency spacing between modes), typically a few nanoseconds or less. In most cases, these lasers are still termed "continuous wave" as their output power is steady when averaged over any longer time periods, with the very high-frequency power variations having little or no impact in the intended application. However, the term is not applied to mode-locked lasers, where the *intention* is to create very short pulses at the rate of the round-trip time.

For continuous wave operation, it is required for the population inversion of the gain medium to be continually replenished by a steady pump source. In some lasing media, this is impossible. In some other lasers, it would require pumping the laser at a very high continuous power level which would be impractical or destroy the laser by producing excessive heat. Such lasers cannot be run in CW mode.

Pulsed Operation

Pulsed operation of lasers refers to any laser not classified as continuous wave, so that the optical power appears in pulses of some duration at some repetition rate. This encompasses a wide range of technologies addressing a number of different motivations. Some lasers are pulsed simply because they cannot be run in continuous mode.

In other cases, the application requires the production of pulses having as large an energy as possible. Since the pulse energy is equal to the average power divided by the repetition rate, this goal can sometimes be satisfied by lowering the rate of pulses so that more energy can be built up in between pulses. In laser ablation, for example, a small volume of material at the surface of a work piece can be evaporated if it is heated in a very short time, while supplying the energy gradually would allow for the heat to be absorbed into the bulk of the piece, never attaining a sufficiently high temperature at a particular point.

Other applications rely on the peak pulse power (rather than the energy in the pulse), especially in order to obtain nonlinear optical effects. For a given pulse energy, this requires creating pulses of the shortest possible duration utilizing techniques such as Q-switching.

The optical bandwidth of a pulse cannot be narrower than the reciprocal of the pulse width. In the case of extremely short pulses, that implies lasing over a considerable bandwidth, quite contrary to the very narrow bandwidths typical of CW lasers. The lasing medium in some *dye lasers* and *vibronic solid-state lasers* produces optical gain over a wide bandwidth, making a laser possible which can thus generate pulses of light as short as a few femtoseconds (10^{-15} s).

Q-switching

In a Q-switched laser, the population inversion is allowed to build up by introducing loss inside the resonator which exceeds the gain of the medium; this can also be described as a reduction of the quality factor or 'Q' of the cavity. Then, after the pump energy stored in the laser medium has approached the maximum possible level, the introduced loss mechanism (often an electro- or acousto-optical element) is rapidly removed (or that occurs by itself in a passive device), allowing lasing to begin which rapidly obtains the stored energy in the gain medium. This results in a short pulse incorporating that energy, and thus a high peak power.

Mode-locking

A mode-locked laser is capable of emitting extremely short pulses on the order of tens of picoseconds down to less than 10 femtoseconds. These pulses will repeat at the round trip time, that is, the time that it takes light to complete one round trip between the mirrors comprising the resonator. Due to the Fourier limit (also known as energy-time uncertainty), a pulse of such short temporal length has a spectrum spread over a considerable bandwidth. Thus such a gain medium must have a gain bandwidth sufficiently broad to amplify those frequencies. An example of a suitable material is titanium-doped, artificially grown sapphire (Ti:sapphire) which has a very wide gain bandwidth and can thus produce pulses of only a few femtoseconds duration.

Such mode-locked lasers are a most versatile tool for researching processes occurring on extremely short time scales (known as femtosecond physics, femtosecond chemistry and ultrafast science), for maximizing the effect of nonlinearity in optical materials (e.g. in second-harmonic generation, parametric down-conversion, optical parametric oscillators and the like). Due to the large peak power and the ability to generate phase-stabilized trains of ultrafast laser pulses, mode-locking ultrafast lasers underpin precision metrology and spectroscopy applications.

Pulsed Pumping

Another method of achieving pulsed laser operation is to pump the laser material with a source that is itself pulsed, either through electronic charging in the case of flash lamps, or another laser which is already pulsed. Pulsed pumping was historically used with dye lasers where the inverted population lifetime of a dye molecule was so short that a high energy, fast pump was needed. The way to overcome this problem was to charge up large capacitors which are then switched to discharge through flashlamps, producing an intense flash. Pulsed pumping is also required for three-level lasers in which the lower energy level rapidly becomes highly populated preventing further lasing until those atoms relax to the ground state. These lasers, such as the excimer laser and the copper vapor laser, can never be operated in CW mode.

Foundations

In 1917, Albert Einstein established the theoretical foundations for the laser and the maser in the paper *Zur Quantentheorie der Strahlung* (On the Quantum Theory of Radiation) via a re-derivation of Max Planck's law of radiation, conceptually based upon probability coefficients (Einstein coefficients) for the absorption, spontaneous emission, and stimulated emission of electromagnetic radiation. In 1928, Rudolf W. Ladenburg confirmed the existence of the phenomena of stimulated emission and negative absorption. In 1939, Valentin A. Fabrikant predicted the use of stimulated emission to amplify "short" waves. In 1947, Willis E. Lamb and R.C. Retherford found apparent stimulated emission in hydrogen spectra and effected the first demonstration of stimulated emission. In 1950, Alfred Kastler proposed the method of optical pumping, experimentally confirmed, two years later, by Brossel, Kastler, and Winter.

Maser

Aleksandr Prokhorov.

In 1951, Joseph Weber submitted a paper on using stimulated emissions to make a microwave amplifier to the June 1952 Institute of Radio Engineers Vacuum Tube Research Conference at Ottawa, Ontario, Canada. After this presentation, RCA asked Weber to give a seminar on this idea, and Charles Hard Townes asked him for a copy of the paper.

In 1953, Charles Hard Townes and graduate students James P. Gordon and Herbert J. Zeiger produced the first microwave amplifier, a device operating on similar principles to the laser, but amplifying microwave radiation rather than infrared or visible radiation. Townes's maser was incapable of continuous output. Meanwhile, in the Soviet Union, Nikolay Basov and Aleksandr Prokhorov were independently working on the quantum oscillator and solved the problem of continuous-output systems by using more than two energy levels. These gain media could release stimulated emissions between an excited state and a lower excited state, not the ground state, facilitating the maintenance of a population inversion. In 1955, Prokhorov and Basov suggested optical pumping of a multi-level system as a method for obtaining the population inversion, later a main method of laser pumping.

Townes reports that several eminent physicists—among them Niels Bohr, John von Neumann, and Llewellyn Thomas—argued the maser violated Heisenberg's uncertainty principle and hence could not work. Others such as Isidor Rabi and Polykarp Kusch expected that it would be impractical and not worth the effort. In 1964 Charles H. Townes, Nikolay Basov, and Aleksandr Prokhorov shared the Nobel Prize in Physics, "for fundamental work in the field of quantum electronics, which has led to the construction of oscillators and amplifiers based on the maser–laser principle".

Laser

In 1957, Charles Hard Townes and Arthur Leonard Schawlow, then at Bell Labs, began a serious study of the infrared laser. As ideas developed, they abandoned infrared radiation to instead concentrate upon visible light. The concept originally was called an "optical maser". In 1958, Bell Labs filed a patent application for their proposed optical maser; and Schawlow and Townes submitted a manuscript of their theoretical calculations to the Physical Review.

Simultaneously, at Columbia University, graduate student Gordon Gould was working on a doctoral thesis about the energy levels of excited thallium. When Gould and Townes met, they spoke of radiation emission, as a general subject; afterwards, in November 1957, Gould noted his ideas for a "laser", including using an open resonator (later an essential laser-device component). Moreover, in 1958, Prokhorov independently proposed using an open resonator, the first published appearance (in the USSR) of this idea. Elsewhere, in the U.S., Schawlow and Townes had agreed to an open-resonator laser design – apparently unaware of Prokhorov's publications and Gould's unpublished laser work.

Gould's linguistic intention was using the "-aser" word particle as a suffix – to accurately denote the spectrum of the light emitted by the LASER device; thus x-rays: *xaser*, ultraviolet: *uvaser*, et cetera; none established itself as a discrete term, although "raser" was briefly popular for denoting radio-frequency-emitting devices.

Gould's notes included possible applications for a laser, such as spectrometry, interferometry, radar, and nuclear fusion. He continued developing the idea, and filed a patent application in April 1959. The U.S. Patent Office denied his application, and awarded a patent to Bell Labs, in 1960. That provoked a twenty-eight-year lawsuit, featuring scientific prestige and money as the stakes. Gould won his first minor patent in 1977, yet it was not until 1987 that he won the first significant patent lawsuit victory, when a Federal judge ordered the U.S. Patent Office to issue patents to Gould for the optically pumped and the gas discharge laser devices. The question of just how to assign credit for inventing the laser remains unresolved by historians.

On May 16, 1960, Theodore H. Maiman operated the first functioning laser at Hughes Research Laboratories, Malibu, California, ahead of several research teams, including those of Townes, at Columbia University, Arthur Schawlow, at Bell Labs, and Gould, at the TRG (Technical Research Group) company. Maiman's functional laser used a flashlamp-pumped synthetic ruby crystal to produce red laser light at 694 nanometers wavelength. The device was only capable of pulsed operation, due to its three-level pumping design scheme. Later that year, the Iranian physicist Ali Javan, and William R. Bennett, and Donald Herriott, constructed the first gas laser, using helium and neon that was capable of continuous operation in the infrared; later, Javan received the Albert Einstein Award in 1993. Basov and Javan proposed the semiconductor laser diode concept. In 1962, Robert N. Hall demonstrated the first *laser diode* device, which was made of gallium arsenide and emitted in the near-infrared band of the spectrum at 850 nm. Later that year, Nick Holonyak, Jr. demonstrated the first semiconductor laser with a visible emission. This first semiconductor laser could only be used in pulsed-beam operation, and when cooled to liquid nitrogen temperatures (77 K). In 1970, Zhores Alferov, in the USSR, and Izuo Hayashi and Morton Panish of Bell Telephone Laboratories also independently developed room-temperature, continual-operation diode lasers, using the heterojunction structure.

Recent Innovations

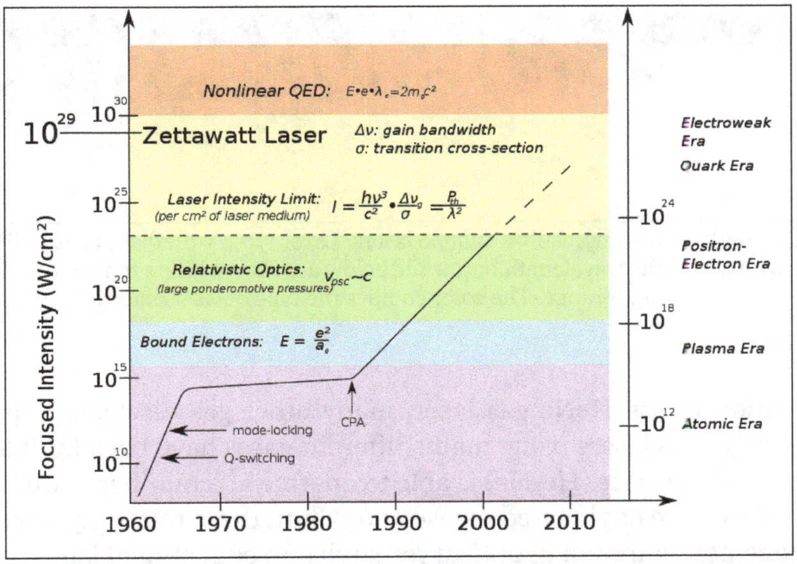

Graph showing the history of maximum laser pulse intensity throughout the past 40 years.

Since the early period of laser history, laser research has produced a variety of improved and specialized laser types, optimized for different performance goals, including:

- New wavelength bands
- Maximum average output power
- Maximum peak pulse energy
- Maximum peak pulse power
- Minimum output pulse duration
- Minimum linewidth
- Maximum power efficiency
- Minimum cost

In 2017, researchers at TU Delft demonstrated an AC Josephson junction microwave laser. Since the laser operates in the superconducting regime, it is more stable than other semiconductor-based lasers. The device has potential for applications in quantum computing. In 2017, researchers at TU Munich demonstrated the smallest mode locking laser capable of emitting pairs of phase-locked picosecond laser pulses with a repetition frequency up to 200 GHz.

In 2017, researchers from the Physikalisch-Technische Bundesanstalt (PTB), together with US researchers from JILA, a joint institute of the National Institute of Standards and Technology (NIST) and the University of Colorado Boulder, established a new world record by developing an erbium-doped fiber laser with a linewidth of only 10 millihertz.

Types and Operating Principles

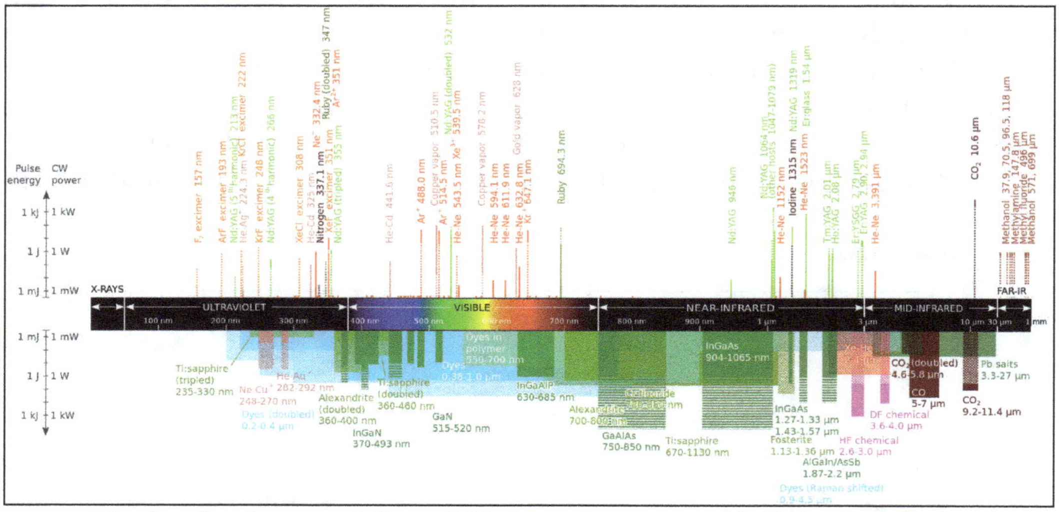

Wavelengths of commercially available lasers. Laser types with distinct laser lines are shown above the wavelength bar, while below are shown lasers that can emit in a wavelength range. The color codifies the type of laser material.

Gas Lasers

Following the invention of the HeNe gas laser, many other gas discharges have been found to amplify light coherently. Gas lasers using many different gases have been built and used for many purposes. The helium–neon laser (He-Ne) is able to operate at a number of different wavelengths, however the vast majority are engineered to lase at 633 nm; these relatively low cost but highly coherent lasers are extremely common in optical research and educational laboratories. Commercial carbon dioxide (CO_2) lasers can emit many hundreds of watts in a single spatial mode which can be concentrated into a tiny spot. This emission is in the thermal infrared at 10.6 μm; such lasers are regularly used in industry for cutting and welding. The efficiency of a CO_2 laser is unusually high: over 30%. Argon-ion lasers can operate at a number of lasing transitions between 351 and 528.7 nm. Depending on the optical design one or more of these transitions can be lasing simultaneously; the most commonly used lines are 458 nm, 488 nm and 514.5 nm. A nitrogen transverse electrical discharge in gas at atmospheric pressure (TEA) laser is an inexpensive gas laser, often home-built by hobbyists, which produces rather incoherent UV light at 337.1 nm. Metal ion lasers are gas lasers that generate deep ultraviolet wavelengths. Helium-silver (HeAg) 224 nm and neon-copper (NeCu) 248 nm are two examples. Like all low-pressure gas lasers, the gain media of these lasers have quite narrow oscillation linewidths, less than 3 GHz (0.5 picometers), making them candidates for use in fluorescence suppressed Raman spectroscopy.

Chemical Lasers

Chemical lasers are powered by a chemical reaction permitting a large amount of energy to be released quickly. Such very high power lasers are especially of interest to the military, however continuous wave chemical lasers at very high power levels, fed by streams of gasses, have been developed and have some industrial applications. As examples, in the hydrogen fluoride laser (2700–2900 nm) and the deuterium fluoride laser (3800 nm) the reaction is the combination of hydrogen or deuterium gas with combustion products of ethylene in nitrogen trifluoride.

Excimer Lasers

Excimer lasers are a special sort of gas laser powered by an electric discharge in which the lasing medium is an excimer, or more precisely an exciplex in existing designs. These are molecules which can only exist with one atom in an excited electronic state. Once the molecule transfers its excitation energy to a photon, its atoms are no longer bound to each other and the molecule disintegrates. This drastically reduces the population of the lower energy state thus greatly facilitating a population inversion. Excimers currently used are all noble gas compounds; noble gasses are chemically inert and can only form compounds while in an excited state. Excimer lasers typically operate at ultraviolet wavelengths with major applications including semiconductor photolithography and LASIK eye surgery. Commonly used excimer molecules include ArF (emission at 193 nm), KrCl (222 nm), KrF (248 nm), XeCl (308 nm), and XeF (351 nm). The molecular fluorine laser, emitting at 157 nm in the vacuum ultraviolet is sometimes referred to as an excimer laser, however this appears to be a misnomer inasmuch as F_2 is a stable compound.

Solid-state Lasers

A 50 W FASOR, based on a Nd:YAG laser, used at the Starfire Optical Range.

Solid-state lasers use a crystalline or glass rod which is "doped" with ions that provide the required energy states. For example, the first working laser was a ruby laser, made from ruby (chromium-doped corundum). The population inversion is actually maintained in the dopant. These materials are pumped optically using a shorter wavelength than the lasing wavelength, often from a flashtube or from another laser. The usage of the term "solid-state" in laser physics is narrower than in typical use. Semiconductor lasers (laser diodes) are typically *not* referred to as solid-state lasers.

Neodymium is a common dopant in various solid-state laser crystals, including yttrium orthovanadate (Nd:YVO_4), yttrium lithium fluoride (Nd:YLF) and yttrium aluminium garnet (Nd:YAG). All these lasers can produce high powers in the infrared spectrum at 1064 nm. They are used for cutting, welding and marking of metals and other materials, and also in spectroscopy and for pumping dye lasers. These lasers are also commonly frequency doubled, tripled or quadrupled to produce 532 nm (green, visible), 355 nm and 266 nm (UV) beams, respectively. Frequency-doubled diode-pumped solid-state (DPSS) lasers are used to make bright green laser pointers.

Ytterbium, holmium, thulium, and erbium are other common "dopants" in solid-state lasers. Ytterbium is used in crystals such as Yb:YAG, Yb:KGW, Yb:KYW, Yb:SYS, Yb:BOYS, Yb:CaF_2, typically operating around 1020–1050 nm. They are potentially very efficient and high powered due to a small quantum defect. Extremely high powers in ultrashort pulses can be achieved with Yb:YAG. Holmium-doped YAG crystals emit at 2097 nm and form an efficient laser operating at infrared wavelengths strongly absorbed by water-bearing tissues. The Ho-YAG is usually operated in a pulsed mode, and passed through optical fiber surgical devices to resurface joints, remove rot from teeth, vaporize cancers, and pulverize kidney and gall stones.

Titanium-doped sapphire (Ti:sapphire) produces a highly tunable infrared laser, commonly used for spectroscopy. It is also notable for use as a mode-locked laser producing ultrashort pulses of extremely high peak power.

Thermal limitations in solid-state lasers arise from unconverted pump power that heats the medium. This heat, when coupled with a high thermo-optic coefficient (dn/dT) can cause thermal lensing and reduce the quantum efficiency. Diode-pumped thin disk lasers overcome these issues by having a gain medium that is much thinner than the diameter of the pump beam. This allows for a more uniform temperature in the material. Thin disk lasers have been shown to produce beams of up to one kilowatt.

Fiber Lasers

Solid-state lasers or laser amplifiers where the light is guided due to the total internal reflection in a single mode optical fiber are instead called fiber lasers. Guiding of light allows extremely long gain regions providing good cooling conditions; fibers have high surface area to volume ratio which allows efficient cooling. In addition, the fiber's waveguiding properties tend to reduce thermal distortion of the beam. Erbium and ytterbium ions are common active species in such lasers.

Quite often, the fiber laser is designed as a double-clad fiber. This type of fiber consists of a fiber core, an inner cladding and an outer cladding. The index of the three concentric layers is chosen so that the fiber core acts as a single-mode fiber for the laser emission while the outer cladding acts as a highly multimode core for the pump laser. This lets the pump propagate a large amount of power into and through the active inner core region, while still having a high numerical aperture (NA) to have easy launching conditions.

Pump light can be used more efficiently by creating a fiber disk laser, or a stack of such lasers.

Fiber lasers have a fundamental limit in that the intensity of the light in the fiber cannot be so high that optical nonlinearities induced by the local electric field strength can become dominant and prevent laser operation and lead to the material destruction of the fiber. This effect is called photodarkening. In bulk laser materials, the cooling is not so efficient, and it is difficult to separate the effects of photodarkening from the thermal effects, but the experiments in fibers show that the photodarkening can be attributed to the formation of long-living color centers.

Photonic Crystal Lasers

Photonic crystal lasers are lasers based on nano-structures that provide the mode confinement and the density of optical states (DOS) structure required for the feedback to take place. They are typical micrometer-sized and tunable on the bands of the photonic crystals.

Semiconductor Lasers

A 5.6 mm 'closed can' commercial laser diode, probably from a CD or DVD player.

Semiconductor lasers are diodes which are electrically pumped. Recombination of electrons and holes created by the applied current introduces optical gain. Reflection from the ends of the crystal form an optical resonator, although the resonator can be external to the semiconductor in some designs.

Commercial laser diodes emit at wavelengths from 375 nm to 3500 nm. Low to medium power laser diodes are used in laser pointers, laser printers and CD/DVD players. Laser diodes are also frequently used to optically pump other lasers with high efficiency. The highest power industrial laser diodes, with power up to 20 kW, are used in industry for cutting and welding. External-cavity semiconductor lasers have a semiconductor active medium in a larger cavity. These devices can generate high power outputs with good beam quality, wavelength-tunable narrow-linewidth radiation, or ultrashort laser pulses.

In 2012, Nichia and OSRAM developed and manufactured commercial high-power green laser diodes (515/520 nm), which compete with traditional diode-pumped solid-state lasers.

Vertical cavity surface-emitting lasers (VCSELs) are semiconductor lasers whose emission direction is perpendicular to the surface of the wafer. VCSEL devices typically have a more circular output beam than conventional laser diodes. As of 2005, only 850 nm VCSELs are widely available, with 1300 nm VCSELs beginning to be commercialized, and 1550 nm devices an area of research. VECSELs are external-cavity VCSELs. Quantum cascade lasers are semiconductor lasers that have an active transition between energy *sub-bands* of an electron in a structure containing several quantum wells.

The development of a silicon laser is important in the field of optical computing. Silicon is the material of choice for integrated circuits, and so electronic and silicon photonic components (such as optical interconnects) could be fabricated on the same chip. Unfortunately, silicon is a difficult lasing material to deal with, since it has certain properties which block lasing. However, recently teams have produced silicon lasers through methods such as fabricating the lasing material from silicon and other semiconductor materials, such as indium(III) phosphide or gallium(III) arsenide, materials which allow coherent light to be produced from silicon. These are called hybrid silicon laser. Recent developments have also shown the use of monolithically integrated nanowire lasers directly on silicon for optical interconnects, paving the way for chip level applications. These heterostructure nanowire lasers capable of optical interconnects in silicon are also capable of emitting pairs of phase-locked picosecond pulses with a repetition frequency up to 200 GHz, allowing for on-chip optical signal processing. Another type is a Raman laser, which takes advantage of Raman scattering to produce a laser from materials such as silicon.

Lasing without maintaining the medium excited into a population inversion was demonstrated in 1992 in sodium gas and again in 1995 in rubidium gas by various international teams. This was accomplished by using an external maser to induce "optical transparency" in the medium by introducing and destructively interfering the ground electron transitions between two paths, so that the likelihood for the ground electrons to absorb any energy has been cancelled.

Dye Lasers

Close-up of a table-top dye laser based on Rhodamine 6G.

Dye lasers use an organic dye as the gain medium. The wide gain spectrum of available dyes, or mixtures of dyes, allows these lasers to be highly tunable, or to produce very short-duration pulses (on the order of a few femtoseconds). Although these tunable lasers are mainly known in their liquid form, researchers have also demonstrated narrow-linewidth tunable emission in dispersive oscillator configurations incorporating solid-state dye gain media. In their most prevalent form these solid state dye lasers use dye-doped polymers as laser media.

Free-electron Lasers

The free-electron laser *FELIX* at the FOM Institute for Plasma Physics Rijnhuizen, Nieuwegein.

Free-electron lasers, or FELs, generate coherent, high power radiation that is widely tunable, currently ranging in wavelength from microwaves through terahertz radiation and infrared to the visible spectrum, to soft X-rays. They have the widest frequency range of any laser type. While FEL beams share the same optical traits as other lasers, such as coherent radiation, FEL operation is quite different. Unlike gas, liquid, or solid-state lasers, which rely on bound atomic or molecular states, FELs use a relativistic electron beam as the lasing medium, hence the term *free-electron*.

Exotic Media

The pursuit of a high-quantum-energy laser using transitions between isomeric states of an atomic nucleus has been the subject of wide-ranging academic research since the early 1970s. This research has been international in scope, but mainly based in the former Soviet Union and the United States. While many scientists remain optimistic that a breakthrough is near, an operational gamma-ray laser is yet to be realized.

Some of the early studies were directed toward short pulses of neutrons exciting the upper isomer state in a solid so the gamma-ray transition could benefit from the line-narrowing of Mössbauer effect. In conjunction, several advantages were expected from two-stage pumping of a three-level system. It was conjectured that the nucleus of an atom, embedded in the near field of a laser-driven coherently-oscillating electron cloud would experience a larger dipole field than that of the driving laser. Furthermore, nonlinearity of the oscillating cloud would produce both spatial and temporal harmonics, so nuclear transitions of higher multipolarity could also be driven at multiples of the laser frequency.

In September 2007, the BBC News reported that there was speculation about the possibility of using positronium annihilation to drive a very powerful gamma ray laser. Dr. David Cassidy of the University of California, Riverside proposed that a single such laser could be used to ignite a nuclear fusion reaction, replacing the banks of hundreds of lasers currently employed in inertial confinement fusion experiments.

Space-based X-ray lasers pumped by a nuclear explosion have also been proposed as antimissile weapons. Such devices would be one-shot weapons.

Living cells have been used to produce laser light. The cells were genetically engineered to produce green fluorescent protein (GFP). The GFP is used as the laser's "gain medium", where light amplification takes place. The cells were then placed between two tiny mirrors, just 20 millionths of a meter across, which acted as the "laser cavity" in which light could bounce many times through the cell. Upon bathing the cell with blue light, it could be seen to emit directed and intense green laser light.

Uses

When lasers were invented in 1960, they were called "a solution looking for a problem". Since then, they have become ubiquitous, finding utility in thousands of highly varied applications in every section of modern society, including consumer electronics, information technology, science, medicine, industry, law enforcement, entertainment, and the military. Fiber-optic communication using lasers is a key technology in modern communications, allowing services such as the Internet.

The first widely noticeable use of lasers was the supermarket barcode scanner, introduced in 1974. The laserdisc player, introduced in 1978, was the first successful consumer product to include a laser but the compact disc player was the first laser-equipped device to become common, beginning in 1982 followed shortly by laser printers.

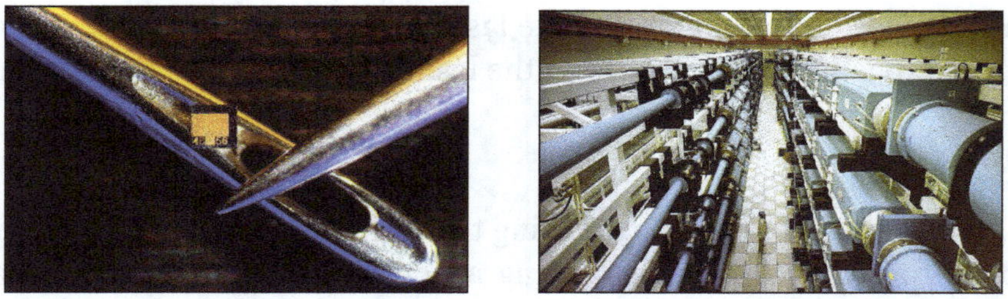

Lasers range in size from microscopic diode lasers (top) with numerous applications, to football field sized neodymium glass lasers (bottom) used for inertial confinement fusion, nuclear weapons research and other high energy density physics experiments.

Some other uses are:

- Communications: Besides fiber-optic communication, lasers are used for free-space optical communication, including laser communication in space.
- Medicine.
- Industry: Cutting including converting thin materials, welding, material heat treatment, marking parts (engraving and bonding), additive manufacturing or 3D printing processes such as selective laser sintering and selective laser melting, non-contact measurement of parts and 3D scanning, and Laser cleaning.
- Military: Marking targets, guiding munitions, missile defense, electro-optical countermeasures (EOCM), lidar, blinding troops.
- Law enforcement: LIDAR traffic enforcement. Lasers are used for latent fingerprint detection in the forensic identification field.
- Research: Spectroscopy, laser ablation, laser annealing, laser scattering, laser interferometry, lidar, laser capture microdissection, fluorescence microscopy, metrology, laser cooling.
- Commercial products: Laser printers, barcode scanners, thermometers, laser pointers, holograms, bubblegrams.
- Entertainment: Optical discs, laser lighting displays, laser turntables.

In 2004, excluding diode lasers, approximately 131,000 lasers were sold with a value of US$2.19 billion. In the same year, approximately 733 million diode lasers, valued at $3.20 billion, were sold.

In Medicine

Lasers have many uses in medicine, including laser surgery (particularly eye surgery), laser healing, kidney stone treatment, ophthalmoscopy, and cosmetic skin treatments such as acne treatment, cellulite and striae reduction, and hair removal.

Lasers are used to treat cancer by shrinking or destroying tumors or precancerous growths. They are most commonly used to treat superficial cancers that are on the surface of the body or the lining of internal organs. They are used to treat basal cell skin cancer and the very early stages of others like cervical, penile, vaginal, vulvar, and non-small cell lung cancer. Laser therapy is often combined with other treatments, such as surgery, chemotherapy, or radiation therapy. Laser-induced interstitial thermotherapy (LITT), or interstitial laser photocoagulation, uses lasers to treat some cancers using hyperthermia, which uses heat to shrink tumors by damaging or killing cancer cells. Lasers are more precise than traditional surgery methods and cause less damage, pain, bleeding, swelling, and scarring. A disadvantage is that surgeons must have specialized training. It may be more expensive than other treatments.

As Weapons

Many types of laser can potentially be used as incapacitating weapons, through their ability to produce temporary or permanent vision loss when aimed at the eyes. The degree, character, and duration of vision impairment caused by eye exposure to laser light varies with the power of the laser, the wavelength(s), the collimation of the beam, the exact orientation of the beam, and the duration of exposure. Lasers of even a fraction of a watt in power can produce immediate, permanent vision loss under certain conditions, making such lasers potential non-lethal but incapacitating weapons. The extreme handicap that laser-induced blindness represents makes the use of lasers even as non-lethal weapons morally controversial, and weapons designed to cause permanent blindness have been banned by the Protocol on Blinding Laser Weapons. Weapons designed to cause temporary blindness, known as dazzlers, are used by military and sometimes law enforcement organizations. Incidents of pilots being exposed to lasers while flying have prompted aviation authorities to implement special procedures to deal with such hazards.

The US-Israeli Tactical High Energy weapon has been used to shoot down rockets and artillery shells.

Laser weapons capable of directly damaging or destroying a target in combat are still in the experimental stage. The general idea of laser-beam weaponry is to hit a target with a train of brief pulses of light. The rapid evaporation and expansion of the surface causes shockwaves that damage the target. The power needed to project a high-powered laser beam of this kind is beyond the limit of current mobile power technology, thus favoring chemically powered gas dynamic lasers. Example experimental systems include MIRACL and the Tactical High Energy Laser.

Throughout the 2000s, the United States Air Force worked on the Boeing YAL-1, an airborne laser mounted in a Boeing 747. It was intended to be used to shoot down incoming ballistic missiles over enemy territory. In March 2009, Northrop Grumman claimed that its engineers in Redondo Beach had successfully built and tested an electrically powered solid state laser capable of producing a 100-kilowatt beam, powerful enough to destroy an airplane. According to Brian Strickland, manager for the United States Army's Joint High Power Solid State Laser program, an electrically powered laser is capable of being mounted in an aircraft, ship, or other vehicle because it requires much less space for its supporting equipment than a chemical laser. However, the source of such a large electrical power in a mobile application remained unclear. Ultimately, the project was deemed to be infeasible, and was cancelled in December 2011, with the Boeing YAL-1 prototype being stored and eventually dismantled.

Boeing YAL-1. The laser system is mounted in a turret attached to the aircraft nose.

The United States Navy is developing a laser weapon referred to as the Laser Weapon System or LaWS.

Hobbies

In recent years, some hobbyists have taken interests in lasers. Lasers used by hobbyists are generally of class IIIa or IIIb (see Safety), although some have made their own class IV types. However, compared to other hobbyists, laser hobbyists are far less common, due to the cost and potential dangers involved. Due to the cost of lasers, some hobbyists use inexpensive means to obtain lasers, such as salvaging laser diodes from broken DVD players (red), Blu-ray players (violet), or even higher power laser diodes from CD or DVD burners.

Hobbyists also have been taking surplus pulsed lasers from retired military applications and modifying them for pulsed holography. Pulsed Ruby and pulsed YAG lasers have been used.

Examples by Power

Laser application in astronomical adaptive optics imaging.

Different applications need lasers with different output powers. Lasers that produce a continuous beam or a series of short pulses can be compared on the basis of their average power. Lasers that produce pulses can also be characterized based on the *peak* power of each pulse. The peak power of a pulsed laser is many orders of magnitude greater than its average power. The average output power is always less than the power consumed.

Table: The continuous or average power required for some uses.

Power	Use
1–5 mW	Laser pointers
5 mW	CD-ROM drive
5–10 mW	DVD player or DVD-ROM drive
100 mW	High-speed CD-RW burner
250 mW	Consumer 16× DVD-R burner
400 mW	Burning through a jewel case including disc within 4 seconds
	DVD 24× dual-layer recording
1 W	Green laser in Holographic Versatile Disc prototype development
1–20 W	Output of the majority of commercially available solid-state lasers used for micro machining
30–100 W	Typical sealed CO_2 surgical lasers
100–3000 W	Typical sealed CO_2 lasers used in industrial laser cutting

Examples of pulsed systems with high peak power:

- 700 TW (700×10^{12} W) – National Ignition Facility, a 192-beam, 1.8-megajoule laser system adjoining a 10-meter-diameter target chamber.
- 1.3 PW (1.3×10^{15} W) – World's most powerful laser as of 1998, located at the Lawrence Livermore Laboratory.

Safety

Even the first laser was recognized as being potentially dangerous. Theodore Maiman characterized the first laser as having a power of one "Gillette" as it could burn through one Gillette razor blade. Today, it is accepted that even low-power lasers with only a few milliwatts of output power can be hazardous to human eyesight when the beam hits the eye directly or after reflection from a shiny surface. At wavelengths which the cornea and the lens can focus well, the coherence and low divergence of laser light means that it can be focused by the eye into an extremely small spot on the retina, resulting in localized burning and permanent damage in seconds or even less time.

European laser warning symbol required for Class 2 lasers and higher.

Lasers are usually labeled with a safety class number, which identifies how dangerous the laser is:

- Class 1 is inherently safe, usually because the light is contained in an enclosure, for example in CD players.
- Class 2 is safe during normal use; the blink reflex of the eye will prevent damage. Usually up to 1 mW power, for example laser pointers.
- Class 3R (formerly IIIa) lasers are usually up to 5 mW and involve a small risk of eye damage within the time of the blink reflex. Staring into such a beam for several seconds is likely to cause damage to a spot on the retina.
- Class 3B can cause immediate eye damage upon exposure.
- Class 4 lasers can burn skin, and in some cases, even scattered light can cause eye and skin damage. Many industrial and scientific lasers are in this class.

The indicated powers are for visible-light, continuous-wave lasers. For pulsed lasers and invisible wavelengths, other power limits apply. People working with class 3B and class 4 lasers can protect their eyes with safety goggles which are designed to absorb light of a particular wavelength.

Infrared lasers with wavelengths longer than about 1.4 micrometers are often referred to as "eye-safe", because the cornea tends to absorb light at these wavelengths, protecting the retina from damage. The label "eye-safe" can be misleading, however, as it applies only to relatively low power continuous wave beams; a high power or Q-switched laser at these wavelengths can burn the cornea, causing severe eye damage, and even moderate power lasers can injure the eye.

Lasers can be a hazard to both civil and military aviation, due to the potential to temporarily distract or blind pilots.

Cameras based on charge-coupled devices may actually be more sensitive to laser damage than biological eyes.

AUTOCOLLIMATOR

An autocollimator is an optical instrument that is used to measure small angles with very high sensitivity. As such, the autocollimator has a wide variety of applications including precision alignment, detection of angular movement, verification of angle standards, and angular monitoring over long periods.

Principles of Operation

The autocollimator projects a beam of collimated light. An external reflector reflects all or part of the beam back into the instrument where the beam is focused and detected by a photodetector. The autocollimator measures the deviation between the emitted beam and the reflected beam. Because the autocollimator uses light to measure angles, it never comes into contact with the test surface.

Digital vs. Visual Autocollimators

Digital Autocollimators

Digital autocollimators use an electronic photodetector to detect the reflected beam. Micro-Radian autocollimators take advantage of the latest detector technology including advanced silicon-based photodetectors and germanium-based detectors. The detector sends a signal which is digitized and processed using proprietary DSP-based electronics. The processing creates a calibrated angular output which is traceable to the Swiss Federal Institute of Metrology (METAS).

Digital autocollimators are available as stand-alone bench top models, as USB-based models designed for use with a PC and as optical-head-only models designed for direct integration into OEM systems. Available outputs include a touchscreen LCD display on the bench top models as well as a combination of USB 2.0, RS-232, and analog signals.

The exceptional accuracy and high resolution of digital autocollimators make them suitable for applications including the calibration of rotary tables, the verification of angle standards, and for remote or long term angular monitoring operations. Additional uses include measurements of flatness or straightness and providing angular feedback in servo-controlled systems.

Digital Autocollimator Diagram

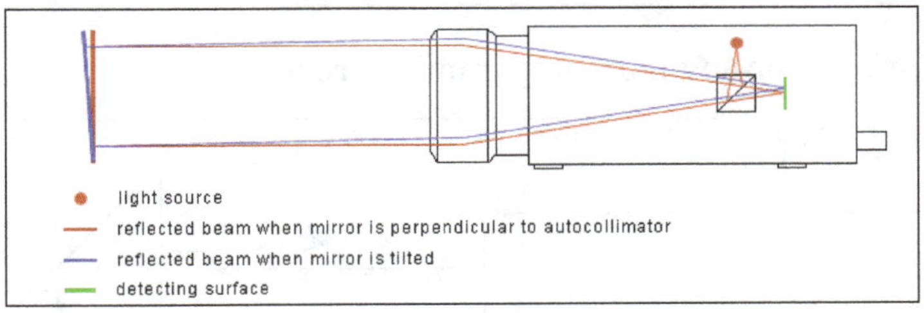

Visual Autocollimators

Visual autocollimators rely on the operator's eye to act as the photodetector. Micro-Radian visual autocollimators project a pinhole image. The operator views the reflected pinhole image(s) through an eyepiece. Because the human eye acts as the photodetector, resolution will vary among operators. Typically, people can resolve from 3 to 5 arc-seconds. Because the human eye is able to discern multiple images simultaneously, visual autocollimators are suitable for measuring multiple surfaces simultaneously. This makes them ideal alignment instruments in applications like aligning laser rod ends or checking parallelism among optics. Visual autocollimators can also be equipped with an eyepiece reticle for aid in lining up test optics to a master reference. All Micro-Radian visual autocollimators are focused on infinity, making them suitable for use at any working distance.

Micro-Radian also offers a visual autocollimator that uses a CCD camera in place of the eyepiece. The camera transmits a video signal via USB 2.0 and the images are viewed on a monitor.

Visual Autocollimator Sample Applications

Measurement of non-parallelism in windows, laser rod ends, and optical wedges by the two-face reflection method.

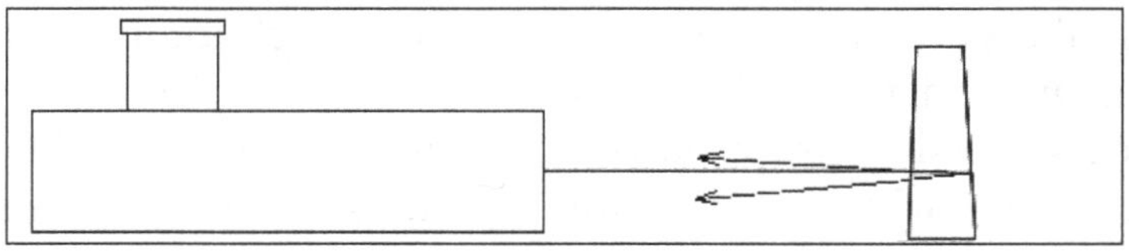

Measurement of squareness of an outside corner by aperture sharing.

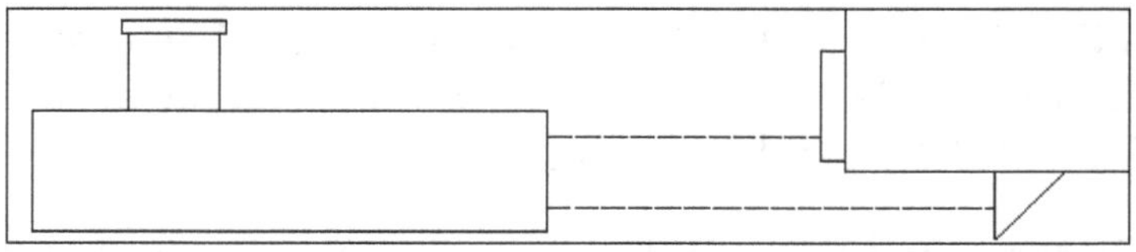

Angle comparisons by aperture sharing.

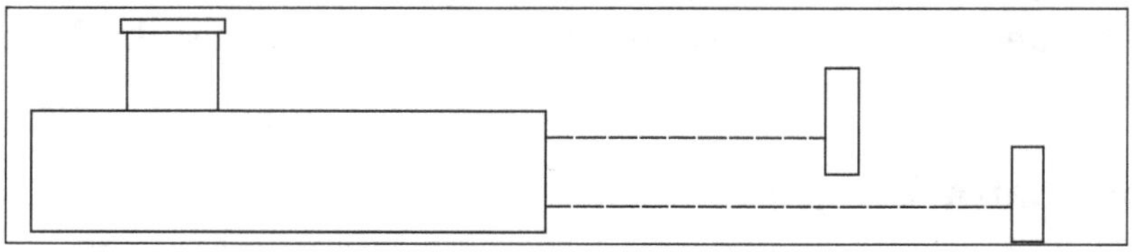

Checking right angle prisms for angular and pyramid errors.

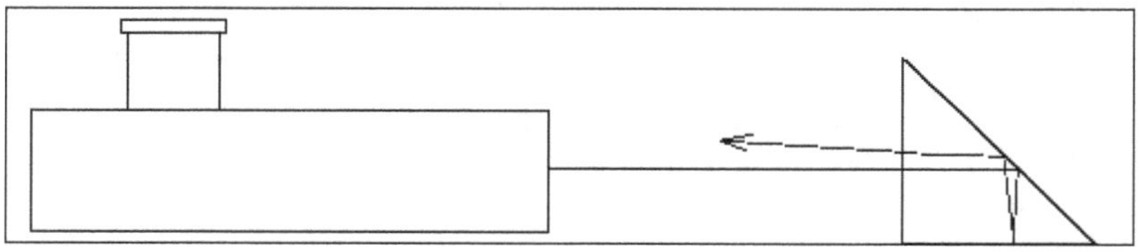

Specialized Autocollimators

Laser Autocollimators

Micro-Radian is the only company that manufactures autocollimators with laser light sources. Micro-Radian laser autocollimators have emitted beams as small as 1.0 mm in diameter. These specialized instruments are designed for measuring small test parts including silicon components, miniature mirrors and lenses, and fiber optic components. They are also ideal any time very low noise measurements or very long distance measurements are desired.

Servo-controlled Systems

Micro-Radian autocollimators are available in both analog or digital "servo" versions suitable for providing high speed, real-time angular monitoring in OEM servo-feedback and servo-feedforward systems. Typically, the servo-version of the T40 or T30 optical heads are selected because of their compact, lightweight size and wide field of view.

References

- Optical-instruments, laboratory-equipment: industrial-needs.com, Retrieved 8 January, 2019
- Thompson, Robert Bruce; Thompson, Barbara Fritchman (2005-06-24). Astronomy Hacks, chapter 1, page 34. ISBN 9780596100605. Retrieved 2009-11-03
- Optical-telescope, science: britannica.com, Retrieved 9 February, 2019
- "The Measurement Shop's Guide to Refractometers - Measurement Shop UK". Www.measurementshop.co.uk. Retrieved 2018-10-15
- Microscopes, technology-terms-and-concepts, technology, science-and-technology: encyclopedia.com, Retrieved 10 March, 2019
- "Daguerreotype". Scientific American. 2 (38): 302. 1847. Doi:10.1038/scientificamerican06121847-302e. ISSN 0036-8733. JSTOR 24924116
- Optical-interferometer, technology: britannica.com, Retrieved 11 April, 2019
- "Self Focusing Binoculars (Fixed Focus): Always in Focus Binoculars". Best Binoculars & Binocular Reviews Website. Retrieved 16 June 2012
- Principles: micro-radian.com, Retrieved 12 May, 2019

Optical Sensors

CHAPTER 4

An optical sensor deals with the conversion of light rays into an electronic signal. The major types of optical sensors are fiber-optic sensor and electro-optical sensor. This chapter closely examines these optical sensors as well as their applications to provide an extensive understanding of the subject.

An optical sensor converts light rays into an electronic signal. The purpose of an optical sensor is to measure a physical quantity of light and, depending on the type of sensor, then translates it into a form that is readable by an integrated measuring device. Optical Sensors are used for contact-less detection, counting or positioning of parts. Optical sensors can be either internal or external. External sensors gather and transmit a required quantity of light, while internal sensors are most often used to measure the bends and other small changes in direction.

The measurands possible by different optical sensors are temperature, velocity liquid level, pressure, displacement (position), vibrations, chemical species, force radiation, pH-value, strain, acoustic field and electric field.

Types of Optical Sensors

There are different kinds of optical sensors, the most common types which we have been using in our real world applications as given below:

- Photoconductive devices used to measure the resistance by converting a change of incident light into a change of resistance.
- The photovoltaic cell (solar cell) converts an amount of incident light into an output voltage.
- The Photodiodes convert an amount of incident light into an output current.

Phototransistors are a type of bipolar transistor where the base-collector junction is exposed to light. This results in the same behavior of a photodiode, but with an internal gain.

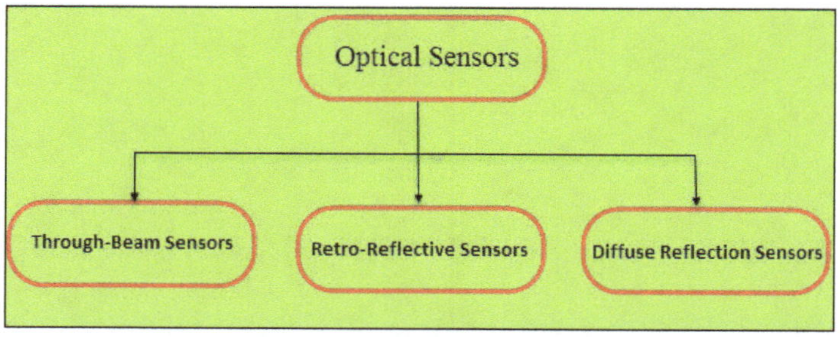

Different Types of Optical Sensors.

The operating principle is the transmitting and receiving of light in an optical sensor, the object to be detected reflects or interrupts a light beam sent out by an emitting diode. Depending on the type of device, the interruption or reflection of the light beam is evaluated. This makes it possible to detect objects independently of the material they are constructed from (wood, metal, plastic or other). Special devices even allow for a detection of transparent objects or those with different colors or variations in contrast. Different types of optical sensors as explained below.

Through-beam Sensors

The system consists of two separate components the transmitter and the receiver are placed opposite to each other. The transmitter projects a light beam onto the receiver. An interruption of the light beam is interpreted as a switch signal by the receiver. It is irrelevant where the interruption occurs.

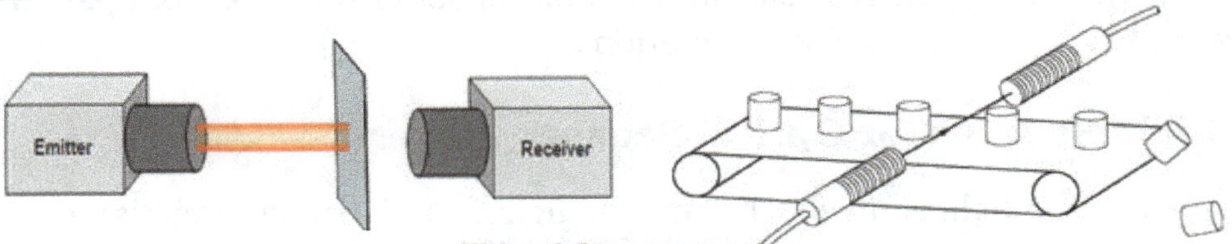

Through-beam sensor

Advantage: Large operating distances can be achieved and the recognition is independent of the object's surface structure, color or reflectivity.

To guarantee a high operational dependability it must be assured that the object is sufficiently large to interrupt the light beam completely.

Retro-reflective Sensors

Transmitter and receiver are both in the same house, through a reflector the emitted light beam is directed back to the receiver. An interruption of the light beam initiates a switching operation. Where the interruption occurs is of no importance.

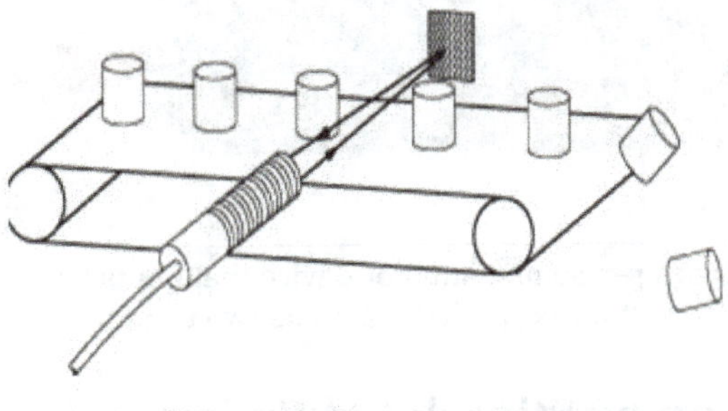

Retro-Reflective Sensors

Advantage: Retro-reflective sensors enable large operating distances with switching points, which are exactly reproducible requiring little mounting effort. All objects interrupting the light beam are accurately detected independently of their surface structure or color.

Diffuse Reflection Sensors

Both transmitter and receiver are in one housing. The transmitted light is reflected by the object to be detected.

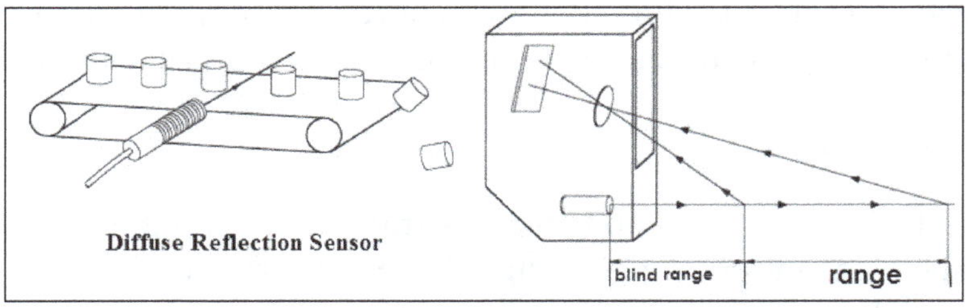

Advantage: The diffused light intensity at the receiver serves as the switching condition. Regardless of the sensitivity setting the rear part always reflects better than the front part. This leads to the consequence to erroneous switching operations.

Different Light Sources for Optical Sensors

There are many types of light sources. The sun and light from burning torch flames were the first light sources used to study optics. As a matter of fact, light coming from certain (exited) matter (e.g., iodine, chlorine, and mercury ions) still provides the reference points in the optical spectrum. One of the key components in optical communication is the monochromatic light source. In optical communications, light sources must be monochromatic, compact, and long lasting. Here are two different types of light source.

Light Emitting Diode

During the recombination process of electrons with holes at the junctions of n-doped and p-doped semiconductors, energy is released in the form of light. The excitation takes place by applying an external voltage and the recombination may be taking place, or it may be stimulated as another photon. This facilitates coupling the LED light with an optical device.

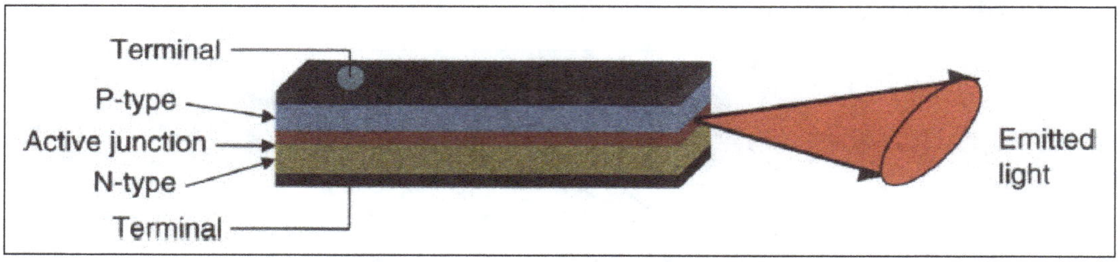

A LED is a p-n semiconductor device that emits light when a voltage is applied across its two terminals.

Laser Light Amplification by Stimulated Emission Radiation

A laser is created, when the electrons in the atoms in special glasses, crystals, or gasses absorb energy from an electrical current they become excited. The excited electrons move from a lower-energy orbit to a higher-energy orbit around the atom's nucleus. When they return to their normal or ground state this leads to the electrons emit photons (particles of light). These photons are all at the same wavelength and coherent. The ordinary visible light comprises multiple wavelengths and is not coherent.

FIBER-OPTIC SENSOR

A fiber-optic sensor is a sensor that uses optical fiber either as the sensing element ("intrinsic sensors"), or as a means of relaying signals from a remote sensor to the electronics that process the signals ("extrinsic sensors"). Fibers have many uses in remote sensing. Depending on the application, fiber may be used because of its small size, or because no electrical power is needed at the remote location, or because many sensors can be multiplexed along the length of a fiber by using light wavelength shift for each sensor, or by sensing the time delay as light passes along the fiber through each sensor. Time delay can be determined using a device such as an optical time-domain reflectometer and wavelength shift can be calculated using an instrument implementing optical frequency domain reflectometry.

Fiber-optic sensors are also immune to electromagnetic interference, and do not conduct electricity so they can be used in places where there is high voltage electricity or flammable material such as jet fuel. Fiber-optic sensors can be designed to withstand high temperatures as well.

Intrinsic Sensors

Optical fibers can be used as sensors to measure strain, temperature, pressure and other quantities by modifying a fiber so that the quantity to be measured modulates the intensity, phase, polarization, wavelength or transit time of light in the fiber. Sensors that vary the intensity of light are the simplest, since only a simple source and detector are required. A particularly useful feature of intrinsic fiber-optic sensors is that they can, if required, provide distributed sensing over very large distances.

Temperature can be measured by using a fiber that has evanescent loss that varies with temperature, or by analyzing the Rayleigh Scattering, Raman scattering or the Brillouin scattering in the optical fiber. Electrical voltage can be sensed by nonlinear optical effects in specially-doped fiber, which alter the polarization of light as a function of voltage or electric field. Angle measurement sensors can be based on the Sagnac effect.

Special fibers like long-period fiber grating (LPG) optical fibers can be used for direction recognition. Photonics Research Group of Aston University in UK has some publications on vectorial bend sensor applications.

Optical fibers are used as hydrophones for seismic and sonar applications. Hydrophone systems with more than one hundred sensors per fiber cable have been developed. Hydrophone sensor systems are used by the oil industry as well as a few countries' navies. Both bottom-mounted hydrophone arrays and towed streamer systems are in use. The German company Sennheiser developed a laser microphone for use with optical fibers.

A fiber-optic microphone and fiber-optic based headphone are useful in areas with strong electrical or magnetic fields, such as communication amongst the team of people working on a patient inside a magnetic resonance imaging (MRI) machine during MRI-guided surgery.

Optical fiber sensors for temperature and pressure have been developed for downhole measurement in oil wells. The fiber-optic sensor is well suited for this environment as it functions at temperatures too high for semiconductor sensors (distributed temperature sensing).

Optical fibers can be made into interferometric sensors such as fiber-optic gyroscopes, which are used in the Boeing 767 and in some car models (for navigation purposes). They are also used to make hydrogen sensors.

Fiber-optic sensors have been developed to measure co-located temperature and strain simultaneously with very high accuracy using fiber Bragg gratings. This is particularly useful when acquiring information from small or complex structures. Fiber Bragg grating sensors are also particularly well suited for remote monitoring, and they can be interrogated 250 km away from the monitoring station using an optical fiber cable. Brillouin scattering effects can also be used to detect strain and temperature over large distances (20–120 kilometers).

Other Examples of Sensors

A fiber-optic AC/DC voltage sensor in the middle and high voltage range (100–2000 V) can be created by inducing measurable amounts of Kerr nonlinearity in single mode optical fiber by exposing a calculated length of fiber to the external electric field. The measurement technique is based on polarimetric detection and high accuracy is achieved in a hostile industrial environment.

High frequency (5 MHz–1 GHz) electromagnetic fields can be detected by induced nonlinear effects in fiber with a suitable structure. The fiber used is designed such that the Faraday and Kerr effects cause considerable phase change in the presence of the external field. With appropriate sensor design, this type of fiber can be used to measure different electrical and magnetic quantities and different internal parameters of fiber material.

Electrical power can be measured in a fiber by using a structured bulk fiber ampere sensor coupled with proper signal processing in a polarimetric detection scheme. Experiments have been carried out in support of the technique.

Fiber-optic sensors are used in electrical switchgear to transmit light from an electrical arc flash to a digital protective relay to enable fast tripping of a breaker to reduce the energy in the arc blast.

Fiber Bragg grating based fiber-optic sensors significantly enhance performance, efficiency and safety in several industries. With FBG integrated technology, sensors can provide detailed analysis and comprehensive reports on insights with very high resolution. These type of sensors are used extensively in several industries like telecommunication, automotive, aerospace, energy, etc. Fiber Bragg gratings are sensitive to the static pressure, mechanical tension and compression and fiber temperature changes. The efficiency of fiber Bragg grating based fiber-optic sensors can be provided by means of central wavelength adjustment of light emitting source in accordance with the current Bragg gratings reflection spectra.

Extrinsic Sensors

Extrinsic fiber-optic sensors use an optical fiber cable, normally a multimode one, to transmit modulated light from either a non-fiber optical sensor, or an electronic sensor connected to an optical transmitter. A major benefit of extrinsic sensors is their ability to reach places which are otherwise inaccessible. An example is the measurement of temperature inside aircraft jet engines by using a fiber to transmit radiation into a radiation pyrometer located outside the engine. Extrinsic sensors can also be used in the same way to measure the internal temperature of electrical transformers, where the extreme electromagnetic fields present make other measurement techniques impossible.

Extrinsic fiber-optic sensors provide excellent protection of measurement signals against noise corruption. Unfortunately, many conventional sensors produce electrical output which must be converted into an optical signal for use with fiber. For example, in the case of a platinum resistance thermometer, the temperature changes are translated into resistance changes. The PRT must therefore have an electrical power supply. The modulated voltage level at the output of the PRT can then be injected into the optical fiber via the usual type of transmitter. This complicates the measurement process and means that low-voltage power cables must be routed to the transducer.

Extrinsic sensors are used to measure vibration, rotation, displacement, velocity, acceleration, torque, and temperature.

Chemical Sensors and Biosensors

It is well-known the propagation of light in optical fiber is confined in the core of the fiber based on the total internal reflection (TIR) principle and near-zero propagation loss within the cladding, which is very important for the optical communication but limits its sensing applications due to the non-interaction of light with surroundings. Therefore, it is essential to exploit novel fiber-optic structures to disturb the light propagation, thereby enabling the interaction of the light with surroundings and constructing fiber-optic sensors. Until now, several methods, including polishing, chemical etching, tapering, bending, as well as femtosecond grating inscription, have been proposed to tailor the light propagation and prompt the interaction of light with sensing materials. In the above-mentioned fiber-optic structures, the enhanced evanescent fields can be efficiently excited to induce the light to expose to and interact with the surrounding medium. However, the fibers themselves can only sense very few kinds of analytes with low-sensitivity and zero-selectivity, which greatly limits their development and applications, especially for biosensors that require both high-sensitivity and high-selectivity. To overcome the issue, an efficient way is to resort to responsive materials, which possess the ability to change their properties, such as RI, absorption, conductivity, etc., once the surrounding environments change. Due to the rapid progress of functional materials in recent years, various sensing materials are available for fiber-optic chemical sensors and biosensors fabrication, including graphene, metals and metal oxides, carbon nanotubes, nanowires, nanoparticles, polymers, quantum dots, etc. Generally, these materials reversibly change their shape/volume upon stimulation by the surrounding environments (the target analysts), which then leads to the variation of RI or absorption of the sensing materials. Consequently, the surrounding changes will be recorded and interrogated by the optical fibers, realizing sensing functions of optical fibers. Currently, various fiber-optic chemical sensors and biosensors have been proposed and demonstrated.

ELECTRO-OPTICAL SENSORS

Electro-optical sensors are electronic detectors that convert light, or a change in light, into an electronic signal. They are used in many industrial and consumer applications, for example:

- Lamps that turn on automatically in response to darkness.
- Position sensors that activate when an object interrupts a light beam.
- Flash detection, to synchronize one photographic flash to another.
- Photoelectric sensors that detect the distance, absence, or presence of an object.

Function: An optical sensor converts light rays into electronic signals. It measures the physical quantity of light and then translates it into a form that is readable by an instrument. An optical sensor is generally part of a larger system that integrates a source of light, a measuring device and the optical sensor. This is often connected to an electrical trigger. The trigger reacts to a change in the signal within the light sensor. An optical sensor can measure the changes from one or several light beams. When a change occurs, the light sensor operates as a photoelectric trigger and therefore either increases or decreases the electrical output. An optical switch enables signals in optical fibres or integrated optical circuits to be switched selectively from one circuit to another. An optical switch can operate by mechanical means or by electro-optic effects, magneto-optic effects as well as by other methods. Optical switches are optoelectronic devices which can be integrated with integrated or discrete microelectronic circuits.

Types of Optical Sensors and Switches

There are many different kinds of optical sensors, the most common types are:

- Photoconductive devices convert a change of incident light into a change of resistance.
- Photovoltaics, commonly known as solar cells, convert an amount of incident light into an output voltage.
- Photodiodes convert an amount of incident light into an output current.
- Phototransistors are a type of bipolar transistor where the base-collector junction is exposed to light. This results in the same behaviour of a photodiode, but with an internal gain.

Optical Switches are usually used in optical fibers, where the electro-optic effect is used to switch one circuit to another. These switches can be implemented with, for example, microelectromechanical systems or piezoelectric systems.

Applications

Electro-optical sensors are used whenever light needs to be converted to energy. Because of this, electro-optical sensors can be seen almost anywhere. Common applications are smartphones where sensors are used to adjust screen brightness, and smartwatches in which sensors are used to measure the wearer's heartbeat.

Optical heart-rate sensor.

Optical sensors can be found in the energy field to monitor structures that generate, produce, distribute, and convert electrical power. The distributed and nonconductive nature of optical fibres makes optical sensors perfect for oil and gas applications, including pipeline monitoring. They can also be found in wind turbine blade monitoring, offshore platform monitoring, power line monitoring and downhole monitoring. Other applications include the civil and transportation fields such as bridge, airport landing strip, dam, railway, airplane, wing, fuel tank and ship hull monitoring.

Among other applications, optical switches can be found in thermal methods which vary the refraction index in one leg of an interferometer in order to switch the signal, MEMS approaches involving arrays of micromirrors that can deflect an optical signal to the appropriate receiver, piezoelectric beam steering liquid crystals which rotate polarized light depending on the applied electric field and acousto-optic methods which change the refraction index as a result of strain induced by an acoustic field to deflect light.

Another important application of optical sensor is to measure the concentration of different compounds by both visible and infrared spectroscopy.

APPLICATIONS OF OPTICAL SENSORS

Application of optical sensors ranges from computers to motion detectors. For optical sensors to work effectively, they must be the correct type for the application, so that they maintain their sensitivity to the property they measure. Some of the common applications include alarm systems, synchros for photographic flashes and systems that can detect the presence of objects.

Applications

Industrial Applications

A variety of in-process inspection and control functions can be performed through non-contact optical sensors. Here are some examples of applications being developed at IMI:

- Temperature measurement by infrared sensors.
- Surface and subsurface detection of defects and delaminations.

- Surface inspection for gauging roughness and thickness.
- Characterization of particle flows (temperature, velocity and diameter).
- In-process composition analysis by laser-induced plasma spectroscopy.
- Real-time thermal imaging of dynamic processes.
- In-situ monitoring of coatings and lubricant films.

Biomedical Applications

Flow monitoring by laser Dopplerimetry is used in several biomedical sensing applications, including dermatology for testing of skin irritants, gastroenterology via endoscopes for making blood perfusion measurements in the stomach and duodenum, etc., and dentistry for contact probes to measure blood flow in the teeth and gums. Sensors are used in internal medicine for angiology and vascular surgery to monitor blood flow during vascular reconstruction and the degree of arteriosclerosis in arteries, and they are used in orthopedics for monitoring the blood perfusion in tissues during and after surgery.

Fibre optic sensors have proven to be ideal transducers for structural monitoring. Being durable, stable and insensitive to external perturbations, they are particularly interesting for the long-term health assessment of civil and geotechnical structures. Many different fibre optic sensor technologies exist and offer a wide range of performances and suitability for different applications. The most widely used sensing techniques include point sensors (Fibre Bragg Gratings and Fabry-Perot interferometers), long-gauge sensors (SOFO) and distributed sensors (Raman and Brillouin scattering sensors). These sensing technologies are now widely used in routine application for health monitoring of structures such as bridges, buildings, monuments, tunnels, dams, dykes, pipelines, landslides and many others. This contribution reviews these systems and technologies and presents some significant application examples, in particular to bridges, buildings, geostructures and pipelines.

Biometrics Applications

Optical sensor scans our finger- it functions like a camera, and obtains our fingerprint image indirectly. We may have similar kind of experience: when a government agency takes our fingerprint, someone will check our fingerprint to see if it is visible. Then, they will put some lotion on our finger, and the operator will press our finger on scanner to take our fingerprint. Optical sensor works in a similar way, so it needs good visible finger, clean finger surface, not too dry.

Applications of Fiber Optic Sensors

Fiber optic sensors are used in several areas. Specifically:

- Measurement of physical properties such as strain,displacement, temperature, pressure, velocity, and acceleration in structures of any shape or size.
- Monitoring the physical health of structures in real time.

- Buildings and Bridges: Concrete monitoring during setting, crack (length, propagation speed) monitoring, prestressing monitoring, spatial displacement measurement, neutral axis evolution, long-term deformation (creep and shrinkage) monitoring, concretesteel interaction, and post-seismic damage evaluation.
- Tunnels: Multipoint optical extensometers, convergence monitoring, concrete / prefabricated vaults evaluation, and joints monitoring damage detection.
- Dams: Foundation monitoring, joint expansion monitoring, spatial displacement measurement, leakage monitoring, and distributed temperature monitoring.
- Heritage structures: Displacement monitoring, crack opening analysis, post-seismic damage evaluation, restoration monitoring, and old-new interaction.

Optical Sensors in Precision Agriculture

Agricultural production, already incredibly efficient in much of the developed world, is increasing each year — this is a necessity to feed the 9+ billion people expected on the planet by 2050. A great example of productivity gains is in corn (i.e., maize, Zea mays). One of the most productive crops, corn's yield has increased sevenfold over the last 100 years in the U.S., meaning that seven times less land is needed to feed the same number of people. While favorable soils and weather have always played a central role in agriculture productivity, the increases in productivity have been driven by farmers' use and adaptation of improved technologies. Despite past successes, agriculture continues to be challenged to "produce more with less" with farm inputs such as phosphorus and irrigation water, which are decreasing in availability and increasing in costs. These challenges are compounded by a changing climate and a need to ensure practices are sustainable both economically and environmentally. As in the past, agriculture will meet new obstacles that arise through research discoveries and new production tools. Unlike in the past, tools will not be exclusively the domain of agronomists, plant breeders, agricultural engineers and other traditional agricultural scientists; finding solutions will require input and integration from diverse disciplines not previously involved in agriculture, such as photonics and computer science. New sensor and imaging technologies are among the most important and exciting new tools that will be incorporated to see and respond to what has previously been intractable. Photonics will help researchers and farmers apply discoveries on the farm to more rapidly, cheaply and accurately manage crops.

Sensors are being incorporated into irrigation systems to apply water based on spectral reflectance.

Quantifying Soil Properties

The land area devoted to agriculture is vast but not homogenous. When zooming from landscape to plant scale at any point in time, important differences emerge that can be difficult to quantify. Sensors are needed to constantly monitor agricultural systems at the highest accuracy feasible. At the coarsest view, the agricultural systems and the specific crops grown in a landscape change across the globe, primarily across latitudes. Agricultural fields within these landscapes, as seen from airplanes and satellites in space, appear as patchworks of different crops adjacent to each other. Adding the dimension of time, specific crops in each field rotate from year to year and biologically reach different stages of a life cycle within a year; each locational and temporal difference results in specific growth requirements to optimize. Zooming in a little closer, there are gradients within fields that have different colors. These gradients are correlated with difference in soil type and abilities of the soil to hold nutrients and moisture. Red, green, blue (RGB), near-infrared spectroscopy (NIRS) and other sensors can quantify these properties throughout the soil profile.

Soil conditions interact with changing weather to impact crop growth; for instance, yield productivity is reduced by flooding in low spots and clay soil in a wet year, but is reduced by drought in high spots and sandy soils in dry years. Soil is formed by different processes over tens of thousands of years and might have bands of sand or clay running through a field in ways that intersect with differences in elevation and microclimate. These and other factors interact to create unique growth conditions for each plant in the field that need to be objectively observed and characterized before they can be managed. While soil amendments to reduce such differences can be used in a personal garden, soil within a farm field or a landscape is impractical to homogenize and must be dealt with as it exists. Unlike the tightly controlled and consistent environments in sensor equipment manufacturing, agricultural production is subject to the biological whims of the crop, the vagaries of the weather and the sovereignty of disconnected markets. Variation must be measured or estimated instead of eliminated — then the crop can be managed for maximum productivity, economic returns and sustainability. This is the objective of precision agriculture and where photonics can play a role.

Modern farming and precision agriculture incorporate a large amount of technology including on tractors, sprayers and combines.

Precision Agriculture

The major goal of precision agriculture is smarter applications of fertilizer, pesticides and irrigation water to maximize efficiency of production and on-farm profit for the plant variety chosen.

Increasingly, these smarter applications have also meant being better stewards of the environment. The over-application of fertilizers or pesticides to areas of a field that do not need them can pollute surface and ground water and cost farmers more per acre, without generating any additional crop yield or quality. Conversely, under-application of inputs reduces overall crop production, which can lead to a financial loss after covering the costs of seed, equipment and land. This delicate balance between over- and under-application differs not just between counties and fields, but can differ incredibly within a field, even plant to plant, and this plant heterogeneity reduces yield.

Nitrogen is among the most widely applied nutrients in agricultural fertilizers and was among the first precision agriculture success stories. Farmers now routinely use multispectral and crop canopy sensors to apply this input more precisely. It has long been known that nitrogen uptake is directly related to the amount of biomass — a normalized difference vegetative index (NDVI = [NIR– red]/[NIR+red]) calculated from reflectance spectra can approximate the "greenness" of the crop, which in turn can indicate crop stress and plant accesses to nitrogen. Active (with their own light source) proximal crop reflectance sensors for NDVI such as Trimble Ag's GreenSeeker or Holland Scientific's Crop Circle have been coupled with variable rate fertilizer controls so that the precise amounts of nitrogen fertilizers can be applied where they are needed in the field. The wavelengths used for NDVI vary by company and sensor and may include other indices than NDVI. In general, red is around 660 nm, NIR is around 770 nm, and red edge is around 735 nm.

Active NDVI proximal crop reflectance sensors, along with ultrasonic and temperature sensors, are mounted on a high-throughput phenotyping vehicle. The same type of NDVI sensors are often attached to a fertilizer rig.

Unmanned Aerial Vehicles

Among new technologies generating the highest interest within agriculture right now are UAVs, fitted with cameras and other imaging sensors (thus becoming unmanned aerial systems, UASs). The value this affords for farmers and crop consultants is in scouting fields for drought stress, pests and disease in a faster and more timely manner, over a larger area, and more systematically than is currently done by manually "walking a field." For most row crops, the goal of scouting is to find aspects of stress that can be managed with inputs. If a farmer has no way to manage a stress, they do not need to know about it and will not be willing to pay for this service. Once stress indicators are identified, the particular stress would require more in-depth manual investigation. Automated sampling methods or higher resolution to diagnose the problem could be available in the future. There are currently two primary types of UAS vehicles being investigated for field-scouting: fixed-wing (plane) aircraft that fly fast and high and copter aircraft that fly slow and low but have much better image resolution. Higher resolution will aid in observation of important things such as which insects and disease lesions are causing issues, which can help to more quickly diagnose the specific problem. However, in addition to being slower and more costly, this also results in more data to have to make decisions from and store.

A UAV copter surveys a field for plant height and shapes at the same time as a lidar (on spray tractor).

A few examples of successful UAS research field problems include the detection of cotton root rot — a nearly intractable soil issue leading to death in cotton — for replanting and fungicide control purposes, the detection of drought yield losses in bean that could be rescued through irrigation and the early detection of a leaf rust epidemic in maize that could be controlled by fungicide. However, despite these early research successes and the fact that UASs can be cheaply purchased and that their use in agriculture is highly publicized, there are so far few examples of UASs creating actionable decisions by farmers. Substantial research and education still needs to be done before this can become a reality. Farmers are finding that there are many steps between taking pictures and extracting usable information, and only some of the steps are commonly known. While the current limiting factors of software, analysis and interpretation of UAV data are solved, new data collection needs will emerge. A major need is for less-expensive and lighter-weight imaging sensors that have faster shutter speed (to reduce pixel smear). Better imaging will further decrease costs and be useful for plant scouting. It should be noted that some companies are already betting that satellites will provide high-enough resolution with frequent-enough revisit time, so that many drawbacks from using UAVs can be avoided.

Comparing leaf rust disease images with visual ratings 1.5 and 8.5
(dead plants are rated a 10) on two varieties of corn in Corpus Christi, Texas, taken from a UAV.

High-throughput Phenotyping

Phenotyping is the act of measuring the physical characteristics of plants and animals. A major goal in plant, animal and human research is to identify genes that control phenotype variation. This requires measuring the DNA and the phenotype of the same organism. While the DNA sequence remains constant, the phenotype can vary dramatically from environment to environment and even within the same field — for example, flood stunting of plants in low spots. The radical improvement in DNA sequencing is one of the few technologies with a rate of advancement exceeding Moore's law for semiconductors and is also partially due to improved imaging. Improvement of phenotype measurement has not kept up and now is often the limiting factor in many genetic studies. High-throughput phenotyping is the scientific community's response, developing faster and cheaper methods of measuring the physical characteristics of a plant. Many of the phenotyping methods emerging involve optical sensors, from simple RGB image sensors to measure plant structures to NIR and Raman spectroscopy point sensors and, increasingly, lidar and multispectral and hyperspectral imaging. Because this research is transdisciplinary and rapidly emerging from disparate disciplines — including basic plant biology, computer vision, engineering and others. Some of the technologies emerging from high-throughput phenotyping will likely be scalable and eventually incorporated into precision agriculture uses. While these phenotyping sensors result in faster and cheaper methods of measurement, they are not always more accurate than manual measurements made by traditional methods. So how important is accuracy in measuring crops?

Approaches in Agricultural Sensing

Agricultural, mechanical and aeronautical engineers often express concerns about accuracy of new sensor and image analysis measurements, but this is less of a concern for biological science interests in phenotyping relative differences, not absolute differences, as long as the results are consistent. Differences in metrics used between engineering and agricultural researchers for assessing success are part of the learning curve of UAS interdisciplinary work. There appear to be two general philosophical approaches that agricultural research and production take with sensor measurements. The first is the comparison with a "gold standard" measurement. For example, comparing the estimates of plant heights from a UAS image with measurements made manually using a ruler, a simple correlation analysis can be conducted to tell if this is a good measurement or not — or can it? If the correlation is good, some will say that the sensor is "as good or better than manual measurements," whereas if the correlation is poor, many will blame the manual measurements and say that they are inaccurate due to "human error," with the assumption that the sensor is more accurate.

In contrast, agricultural research can take a different approach that does not require having standard measurements from which to compare. From an agricultural field scientist's perspective, the first step to evaluate a new photonics sensor estimate may not be to do a correlation but instead determine if the difference between varieties or treatments (drought, disease, etc.) are greater than the difference from measurement error. If so, this can be monitored across years and environments for relationships to yield and sustainable production. This is done in most experiments by including multiple replicates within a field and running restricted maximum likelihood (REML) models to compare how consistent different replications of the same variety and treatment are compared with the error (unexplained) variation between replicates. The measurements that have the least unexplained error variation are the more useful. This approach has found that some days the manual ruler measurements are more consistent for plant height while on other days the UAS sensor estimates are more consistent (it is not known why, but it suggests an interesting line of future research). So while it is tempting to envision sensors as just automating manual measurements that are already made, it is more exciting to envision how photonics could predict crop aspects impractical or impossible to have considered previously — for instance, if a reflectance wavelength can predict a specific disease or deficiency previously unknown. A slight variation on this approach is starting to be used with machine learning, one of the most exciting new analysis methods to be applied to these massive agricultural photonics datasets.

What Types of Sensors are Needed?

There are two main niches of desirable new sensors in the short term for field-based agriculture: improved versions of existing sensors and novel research sensors that can be deployed to measure new phenotypes. First, improved versions of existing sensors should be able to consistently and quickly collect higher-resolution RGB (or preferably multispectral for NDVI) images over increasingly larger areas, being both ruggedized and inexpensive (UAVs frequently crash). Improvements from physically larger sensors, such as better resolution and faster shutter speeds, would allow higher and faster UAS flights for precision agriculture and perhaps eventually satellites. For research phenotyping use of existing sensors, we would like to fly over 500 feet, faster than 30 miles per hour and still get sub-centimeter accuracy on the ground. No sensor to date has this capability, and this resolution is far more than precision agriculture currently requires. Improved sensors will allow agriculture to pixelaborate crops at a level of granularity never before feasible and reach the goal of individual plant management.

Fixed-wing UAVs are more appropriate for surveying farmers' fields than rotary wings because they can fly faster and higher, allowing more land to be covered for lower cost.

For developing novel sensors, extensions of Raman spectroscopy, x-ray fluorescence and hyperspectral spectroscopy across the electromagnetic spectrum would be desirable to detect new compounds from a UAS, but these still would need to be rugged, small enough and with low power requirements to be deployed on a UAV. Sensors could be tuned to detect important hormone or signaling compounds that could indicate stress before other physiological signs are observed, or, more excitingly, could use an open-architecture approach to first find repeatable measures then determine if these never-before-detected signatures correspond to something useful. Research is also starting to look below ground, using technologies like ground-penetrating radar, magnetic resonance imaging, thermoacoustic imaging and x-ray to measure roots. It is critical to recall that it is less important to know what is being measured with these sensors than to demonstrate that the measures are repeatable. Therefore, nearly any new sensor is worth trying in a crop system to identify new properties to be measured in plants.

Challenges of Disrupting Agriculture

A quadcopter flies above crops to image them in Corpus Christi, Texas.
Many different types of imaging sensors are being tested on UAVs.

With many opportunities to apply technology to agriculture, it has recently become fashionable for many companies, both established but especially technology startups, to discuss technology's ability to "disrupt agriculture." It is tempting to naively look at the number of acres involved in agriculture and multiply that by a very small cost and envision very large profits. However, time and time again it has been demonstrated that a real value must be generated for farmers to adopt a new technology regardless of how small the cost, and even when profitable, they still may not adopt it. There are also many unique business challenges to working in agriculture, such as only having one season of each crop per year, uncontrollable experimental losses from weather, the need for a counter-season business strategy when there is no crop in the ground and a regionalism in crops that prevents scaling. For anyone interested in applying a new technology — photonics and otherwise — it is strongly suggested to first find agricultural experts in the area with whom to collaborate. Land-grant universities are a great place to start because they typically have basic, applied, translational and extension research experiments occurring simultaneously that can easily leverage testing of any nondestructive sensor technology.

References

- Strong, Andrew P.; Lees, Gareth; Hartog, Arthur H.; Twohig, Richard; Kader, Kamal; Hilton, Graeme (December 2009). "An Integrated System for Pipeline Condition Monitoring". International Petroleum Technology Conference. Doi:10.2523/IPTC-13661-MS
- Optical-sensors-types-basics-and-applications: elprocus.com, Retrieved 12 June, 2019
- Roth, Wolf-Dieter (2005-04-18). "Der Glasfaser-Schallwandler". Heise Online (in German). Archived from the original on 2008-12-07. Retrieved 2008-07-04
- Optical-Sensors-Advancing-Precision-in: photonics.com, Retrieved 13 July, 2019
- Ghosh, S.K.; Sarkar, S.K.; Chakraborty, S. (2006). "A proposal for single mode fiber optic watt measurement scheme". Journal of Optics (Calcutta). 35 (2): 118–124. Doi:10.1007/BF03354801. ISSN 0972-8821

Optical Communication

CHAPTER 5

Optical communication uses optical fibers to carry signals at a distance via light. Some of its types are fiber-optic communication, optical wireless communication, free-space optical communication and visible light communication. The topics elaborated in this chapter will help in gaining a better understanding of these types of optical communication.

Optical communication is any type of communication in which light is used to carry the signal to the remote end, instead of electrical current. Optical communication relies on optical fibers to carry signals to their destinations. A modulator/demodulator, a transmitter/receiver, a light signal and a transparent channel are the building blocks of the optical communications system.

Because of its numerous advantages over electrical transmission, optical fibers have largely replaced copper wire communications in core networks in the developed world.

Since the development of low-loss optical fiber cables in the 1970s, optical communications became one of the most popular methods of communication. Optical communication systems consist of the following components:

- Transmitter: Converts and transmits an electronic signal into a light signal. The most commonly used transmitters are semiconductor devices, such as light-emitting diodes (LEDs) and laser diodes.
- Receivers: Typically consist of a photo-detector, which converts light into electricity using the photoelectric effect. The photo detector is typically a semiconductor-based photodiode.
- Optical Fiber: Consists of a core, cladding and a buffer through which the cladding guides the light along the core by using total internal reflection.

The main benefits of optical communication include high bandwidth, exceptionally low loss, great transmission range and no electromagnetic interference. The cons of optic communication include the high cost of cable, transmitter/receiver and other support equipment, and the skill and expertise required during cable installation and interconnection.

FIBER-OPTIC COMMUNICATION

Fibre-optic communication has revolutionised the telecommunications industry. It has also made its presence widely felt within the data networking community as well. Using fibre-optic cable, optical communications have enabled telecommunications links to be made over much greater distances and with much lower levels of loss in the transmission medium and possibly most important of all, fiber optical communications has enabled much higher data rates to be accommodated.

As a result of these advantages, fibre-optic communications systems are widely employed for applications ranging from major telecommunications backbone infrastructure to Ethernet systems, broadband distribution, and general data networking.

Development of Fibre-optics

Since the earliest days of telecommunications there has been an ever increasing need to transmit more data even faster. Initially single line wires were used. These gave way to coaxial cables that enabled several channels to transmitted over the same cable. However these systems were limited in bandwidth and optical systems were investigated.

Optical communications became a possibility after the first lasers were developed in the 1960s. The next piece of the jigsaw fell into place when the first optical fibers with a sufficiently low loss for communications purposes were developed in the 1970s. Then, during the late 1970s a considerable amount of research was undertaken. This resulted in the installation of the first optical fibre telecommunications system. It ran over a distance of 45 km and used a wavelength of 0.5 mm and had a data rate of just 45 Mbps - a fraction of what is possible today.

Since then, considerable improvements have been made in the technology. Data rates have improved and in addition to this the performance of the optical fibre has been improved to enable much greater distances to be achieved between repeaters. As an indication of this the speeds that can now be achieved along through a fibre-optic system exceed 10 Tbps.

When the first fibre-optic transmission systems were being developed, it was thought that the fibre-optic cabling and technology would be prohibitively expensive. However, this has not been the case and costs have fallen to the extent that fibre-optics now provides the only viable option for many telecommunications applications. In addition to this it is also used in many local area networks where speed is a major requirement.

Advantages of Fibre-optics for Communications

There are a number of compelling reasons that lead to the widespread adoption of fibre-optic cabling for telecommunications applications:

- Much lower levels of signal attenuation.
- Fibre-optic cabling provides a much higher bandwidth allowing more data to be delivered.
- Fibre-optic cables are much lighter than the coaxial cables that might otherwise be used.
- Fibre-optics do not suffer from stray interference pickup that occurs with coaxial cabling.

Fibre-optic Transmission System

Any fibre-optic data transmission system will comprise a number of different elements. There are three major elements and a further one that is vital for practical systems:

- Transmitter (light source).
- Fibre-optic cable.
- Optical repeater.
- Receiver (Detector).

The different elements of the system will vary according to the application. Systems used for lower capacity links, possibly for local area networks will employ somewhat different techniques and components to those used by network providers that provide extremely high data rates over long distances. Nevertheless the basic principles are the same whatever the system.

In the system the transmitter of light source generates a light stream modulated to enable it to carry the data. Conventionally a pulse of light indicates a "1" and the absence of light indicates "0". This light is transmitted down a very thin fibre of glass or other suitable material to be presented at the receiver or detector. The detector converts the pulses of light into equivalent electrical pulses. In this way the data can be transmitted as light over great distances.

Fibre-optic Transmitter

Although the original telecommunications fibre-optic systems would have used large lasers, today a variety of semiconductor devices can be used. The most commonly used devices are light emitting diodes, LEDs, and semiconductor laser diodes.

The simplest transmitter device is the LED. Its main advantage is that it is cheap, and this makes it ideal for low cost applications where only short runs are needed. However they have a number of drawbacks. The first is that they offer a very low level of efficiency. Only about 1% of the input power enters the optical fibre, and this means that high power drivers would be needed to provide sufficient light to enable long distance transmissions to be made. The other disadvantage of LEDs is that they produce what is termed incoherent light that covers a relatively wide spectrum. Typically the spectral width is between 30 and 60 nm. This means that any chromatic dispersion in the fibre will limit the bandwidth of the system.

In view of their performance, LEDs are used mainly in local-area-network applications where the data rates are typically in the range 10-100 Mb/s and transmission distances are a few kilometres.

Where higher levels of performance are required, i.e. it is necessary that the fibre-optic link can operate over greater distances and with higher data rates, then lasers are used. Although more costly, they offer some significant advantages. In the first instance they are able to provide a higher output level, and in addition to this the light output is directional and this enables a much higher level of efficiency in the transfer of the light into the fibre-optic cable. Typically the coupling efficiency into a single mode fibre may be as high as 50%. A further advantage is that lasers have a very narrow spectral bandwidth as a result of the fact that they produce coherent light. This narrow spectral width enables the lasers to transmit data at much higher rates because modal dispersion is less apparent. Another advantage is that semiconductor lasers can be modulated directly at high frequencies because of short recombination time for the carriers within the semiconductor material.

Laser diodes are often directly modulated. This provides a very simple and effective method of transferring the data onto the optical signal. This is achieved by controlling current applied directly to the device. This in turn varies the light output from the laser. However for very high data rates or very long distance links, it is more effective to run the laser at a constant output level (continuous wave). The light is then modulated using an external device. The advantage of using an external means of modulation is that it increases the maximum link distance because an effect known as laser chirp is eliminated. This chirp broadens the spectrum of the light signal and this increases the chromatic dispersion in the fibre-optic cable.

Fibre-optic Cable

In essence a fibre-optic cable consists of core, around which is another layer referred to as the cladding. Outside of this there is a protective outer coating.

The fibre-optic cables operate because their cladding has a refractive index that is slightly lower than that of the core. This means that light passing down the core undergoes total internal reflection when it reaches the core/cladding boundary, and it is thereby contained within the core of the optical fibre.

Repeaters and Amplifiers

There is a maximum distance over which signals may be transmitted over fibre-optic cabling. This is limited not only by the attenuation of the cable, but also the distortion of the light signal along the cable. In order to overcome these effects and transmit the signals over longer distances, repeaters and amplifiers are used.

Opto-electric repeaters may be used. These devices convert the optical signal into an electrical format where it can be processed to ensure that the signal is not distorted and then converted back into the optical format. It may then be transmitted along the next state of the fibre-optic cable.

An alternative approach is to use an optical amplifier. These amplifiers directly amplify the optical signal without the need to convert the signal back into an electrical format. The amplifiers consist of a length of fibre-optic cable that is doped with a rare earth mineral named Erbium. The treated fibre cable is then illuminated or pumped with light of a shorter wavelength from another laser and this serves to amplify the signal that is being carried.

In view of the much reduced cost of fibre-optic amplifiers over repeaters, amplifiers are far more widely used. Most repeaters have been replaced, and amplifiers are used in virtually all new installations these days.

Receivers

Light travelling along a fibre-optic cable needs to be converted into an electrical signal so that it can be processed and the data that is carried can be extracted. The component that is at the heart of the receiver is a photo-detector. This is normally a semiconductor device and may be a p-n junction, a p-i-n photo-diode or an avalanche photo-diode. Photo-transistors are not used because they do not have sufficient speed.

Once the optical signal from the fibre-optic cable has been applied to the photo-detector and converted into an electrical format it can be processed to recover the data which can then be passed to its final destination.

OPTICAL WIRELESS COMMUNICATION

Optical wireless communications (OWC) is a form of optical communication in which unguided visible, infrared (IR), or ultraviolet (UV) light is used to carry a signal.

OWC systems operating in the visible band (390–750 nm) are commonly referred to as visible light communication (VLC). VLC systems take advantage of light emitting diodes (LEDs) which can be pulsed at very high speeds without noticeable effect on the lighting output and human eye. VLC can be possibly used in a wide range of applications including wireless local area networks, wireless personal area networks and vehicular networks among others. On the other hand, terrestrial point-to-point OWC systems, also known as the free space optical (FSO) systems, operate at the near IR frequencies (750–1600 nm). These systems typically use laser transmitters and offer a cost-effective protocol-transparent link with high data rates, i.e., 10 Gbit/s per wavelength, and provide a potential solution for the backhaul bottleneck. There has also been a growing interest on ultraviolet communication (UVC) as a result of recent progress in solid state optical sources/detectors operating within solar-blind UV spectrum (200–280 nm). In this so-called deep UV band, solar radiation is negligible at the ground level and this makes possible the design of photon-counting detectors with wide field-of-view receivers that increase the received energy with little additional background noise. Such designs are particularly useful for outdoor non-line-of-sight configurations to support low power short-range UVC such as in wireless sensor and ad-hoc networks.

Wireless communications technologies proliferated and became essential very quickly during the last few decades of the 20th century, and the early 21st century. The wide-scale deployment of radio-frequency technologies was a key factor in the expansion of wireless devices and systems. However, the portion of the electromagnetic spectrum used by wireless systems is limited in capacity, and licenses to use parts of the spectrum are expensive. With the rise in data-heavy wireless communications, the demand for RF spectrum is outstripping supply, causing companies to consider options for using parts of the electromagnetic spectrum other than radio frequencies.

Optical wireless communication (OWC) refers to transmission in unguided propagation media through the use of optical carriers: visible, infrared (IR), and ultraviolet (UV) radiation. Signalling through beacon fires, smoke, ship flags and semaphore telegraph can be considered the historical forms of OWC. Sunlight has also been used for long-distance signaling since very early times. The earliest use of sunlight for communication purposes is attributed to ancient Greeks and Romans who used polished shields to send signals by reflecting sunlight during battles. In 1810, Carl Friedrich Gauss invented the heliograph which uses a pair of mirrors to direct a controlled beam of sunlight to a distant station. Although the original heliograph was designed for the geodetic survey, it was used extensively for military purposes during the late 19th and early 20th century. In 1880, Alexander Graham Bell invented the photophone, the world's first wireless telephone system.

Military interest in photophones continued after Bell's time. For example in 1935, the German Army developed a photophone where a tungsten filament lamp with an IR transmitting filter was used as a light source. Also, American and German military laboratories continued the development of high-pressure arc lamps for optical communication until the 1950s. Modern OWC uses either lasers or light emitting diodes (LEDs) as transmitters. In 1962, MIT Lincoln Labs built an experimental OWC link using a light emitting GaAs diode and was able to transmit TV signals over a distance of 30 miles. After the invention of the laser, OWC was envisioned to be the main deployment area for lasers and many trials were conducted using different types of lasers and modulation schemes. However, the results were in general disappointing due to large divergence of laser beams and the inability to cope with atmospheric effects. With the development of low-loss fiber optics in the 1970s, they became the obvious choice for long distance optical transmission and shifted the focus away from OWC systems.

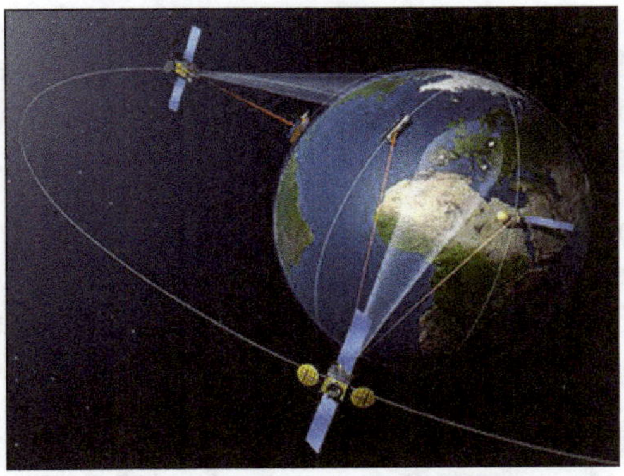

Example of possible ultra-long OWC on inter-satellite European Data Relay Satellite (EDRS) system.

Over the decades, interest in OWC was mainly limited to covert military applications, and space applications including inter-satellite and deep-space links. OWC's mass market penetration has been so far limited with the exception of IrDA which is a highly successful wireless short-range transmission solution.

Applications

Variations of OWC can be potentially employed in a diverse range of communication applications ranging from optical interconnects within integrated circuits through outdoor inter-building links to satellite communications.

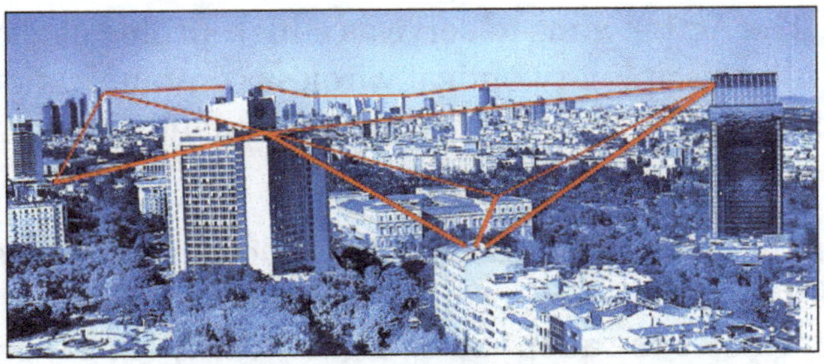

OWC long range inter-buildings communications idea for Istanbul Skyline.

OWC can be divided into five categories based on the transmission range:

- Ultra-short range: Chip-to-chip communications in stacked and closely packed multi-chip packages.
- Short range: Wireless body area network (WBAN) and wireless personal area network (WPAN) applications under standard IEEE 802.15.7, underwater communications.
- Medium range: Indoor IR and visible light communications (VLC) for wireless local area networks (WLANs) and inter-vehicular and vehicle-to-infrastructure communications.
- Long range: Inter-building connections, also called free-space optical communications (FSO).
- Ultra-long range: Laser communication in space especially for inter-satellite links and establishment of satellite constellations.

Recent Trends

- In January 2015, IEEE 802.15 formed a Task Group to write a revision to IEEE 802.15.7-2011 that accommodates infrared and near ultraviolet wavelengths, in addition to visible light, and adds options such as Optical Camera Communications and LiFi.
- At long range OWC applications a 1 Gbit/s - 60 km range link between ground to aircraft at 800 km/h speed has been demonstrated, "Extreme Test for the ViaLight Laser Communication Terminal MLT-20 – Optical Downlink from a Jet Aircraft at 800 km/h", DLR and EADS December 2013.
- On consumer devices and short-range OWC applications on phones; Charge and receive data with light at your smartphone: TCL Communication/ALCATEL ONETOUCH and Sunpartner Technologies announces the first fully integrated solar smartphone. March 2014.
- On ultra-long range OWC applications the NASA's Lunar Laser Communication Demonstration (LLCD) transmitted data from lunar orbit to Earth at a rate of 622 Megabits-per-second (Mbps), November 2013.
- The Next Generation of OWC/Visible Light Communications demonstrated 10 Mbit/s transmission with Polymer Light-Emitting Diodes or OLED.
- On OWC research activities there is a European research project action IC1101 OPTICWISE of the COST Programme (European Cooperation in Science and Technology) funded by the European Science Foundation, allowing the coordination of nationally funded research on a European level. The Action aims to serve as a high-profile consolidated European scientific platform for interdisciplinary optical wireless communication (OWC) research activities. It was launched in November 2011 and will run until November 2015. More than 20 countries represented.
- The consumer and industry OWC technologies adoption is represented by the Li-Fi Consortium, founded in 2011 is a Non-profit organization, devoted to introduce optical wireless technology. Promotes the adoption of Light Fidelity (Li-Fi) products.

- An example of Asian awareness about OWC is the VLCC visible light communication consortium in Japan, established at 2007 in order to realize safe, ubiquitous telecommunication system using visible light through the activities of market research, promotion, and standardization.
- In the USA there are several OWC initiatives, including the "Smart Lighting Engineering Research Center", founded in 2008 by the National Science Foundation (NSF) is a partnership of Rensselaer Polytechnic Institute (lead institution), Boston University and the University of New Mexico. Outreach partners are Howard University, Morgan State University, and Rose-Hulman Institute of Technology.

FREE-SPACE OPTICAL COMMUNICATION

Free-space optical communication (FSO) is an optical communication technology that uses light propagating in free space to wirelessly transmit data for telecommunications or computer networking. "Free space" means air, outer space, vacuum, or something similar. This contrasts with using solids such as optical fiber cable.

An 8-beam free-space optics laser link, rated for 1 Gbit/s. The receptor is the large disc in the middle, the transmitters the smaller ones. At the top right corner is a monocular for assisting the alignment of the two heads.

The technology is useful where the physical connections are impractical due to high costs or other considerations.

Usage and Technologies

Free-space point-to-point optical links can be implemented using infrared laser light, although low-data-rate communication over short distances is possible using LEDs. Infrared Data Association (IrDA) technology is a very simple form of free-space optical communications. On the communications side the FSO technology is considered as a part of the optical wireless communications applications. Free-space optics can be used for communications between spacecraft.

Commercial Products

- In 2008, MRV Communications introduced a free-space optics (FSO)-based system with a data rate of 10 Gbit/s initially claiming a distance of 2 km at high availability. This equipment is no longer available; before end-of-life, the product's useful distance was changed down to 350 m.

- In 2013, the company MOSTCOM started to serially produce a new wireless communication system that also had a data rate of 10 Gbit/s as well as an improved range of up to 2.5 km, but to get to 99.99% uptime the designers used an RF hybrid solution, meaning the data rate drops to extremely low levels during atmospheric disturbances (typically down to 10 Mbit/s). In April 2014, the company with Scientific and Technological Centre "Fiord" demonstrated the transmission speed 30 Gbit/s under "laboratory conditions".
- LightPointe offers many similar hybrid solutions to MOSTCOM's offering.

Useful Distances

The reliability of FSO units has always been a problem for commercial telecommunications. Consistently, studies find too many dropped packets and signal errors over small ranges (400 to 500 meters). This is from both independent studies, such as in the Czech republic, as well as formal internal nationwide studies, such as one conducted by MRV FSO staff. Military based studies consistently produce longer estimates for reliability, projecting the maximum range for terrestrial links is of the order of 2 to 3 km (1.2 to 1.9 mi). All studies agree the stability and quality of the link is highly dependent on atmospheric factors such as rain, fog, dust and heat. Relays may be employed to extend the range for FSO communications.

Extending the Useful Distance

DARPA ORCA official concept art created.

The main reason terrestrial communications have been limited to non-commercial telecommunications functions is fog. Fog consistently keeps FSO laser links over 500 meters from achieving a year-round bit error rate of 1 per 100,000. Several entities are continually attempting to overcome these key disadvantages to FSO communications and field a system with a better quality of service. DARPA has sponsored over US$130 million in research towards this effort, with the ORCA and ORCLE programs.

Other non-government groups are fielding tests to evaluate different technologies that some claim have the ability to address key FSO adoption challenges. As of October 2014, none have fielded a working system that addresses the most common atmospheric events.

FSO research from 1998–2006 in the private sector totaled $407.1 million, divided primarily among four start-up companies. All four failed to deliver products that would meet telecommunications quality and distance standards:

- Terabeam received approximately $575 million in funding from investors such as Softbank, Mobius Venture Capital and Oakhill Venture Partners. AT&T and Lucent backed this attempt. The work ultimately failed, and the company was purchased in 2004 for $52 million by Falls Church, Va.-based YDI, effective June 22, 2004, and used the name Terabeam for the new entity. On September 4, 2007, Terabeam announced it would change its name to Proxim Wireless Corporation, and change its NASDAQ stock symbol from TRBM to PRXM.
- AirFiber received $96.1 million in funding, and never solved the weather issue. They sold out to MRV communications in 2003, and MRV sold their FSO units until 2012 when the end-of-life was abruptly announced for the Terescope series.
- LightPointe Communications received $76 million in start-up funds, and eventually reorganized to sell hybrid FSO-RF units to overcome the weather-based challenges.
- The Maxima Corporation published its operating theory in Science (magazine), and received $9 million in funding before permanently shutting down. No known spin-off or purchase followed this effort.
- Wireless Excellence developed and launched CableFree UNITY solutions that combine FSO with millimeter wave and radio technologies to extend distance, capacity and availability, with a goal of making FSO a more useful and practical technology.

One private company published a paper on November 20, 2014, claiming they had achieved commercial reliability (99.999% availability) in extreme fog. There is no indication this product is currently commercially available.

Extraterrestrial

The massive advantages of laser communication in space have multiple space agencies racing to develop a stable space communication platform, with many significant demonstrations and achievements.

Operational Systems

The first gigabit laser-based communication was achieved by the European Space Agency and called the European Data Relay System (EDRS) on November 28, 2014. The system is operational and is being used on a daily basis.

Demonstrations

NASA's OPALS announced a breakthrough in space-to-ground communication December 9, 2014, uploading 175 megabytes in 3.5 seconds. Their system is also able to re-acquire tracking after the signal was lost due to cloud cover.

In the early morning hours of Oct. 18, NASA's Lunar Laser Communication Demonstration (LLCD) made history, transmitting data from lunar orbit to Earth at a rate of 622 Megabits-per-second (Mbps). LLCD was flown aboard the Lunar Atmosphere and Dust Environment Explorer satellite (LADEE), whose primary science mission was to investigate the tenuous and exotic atmosphere that exists around the moon.

In January 2013, NASA used lasers to beam an image of the Mona Lisa to the Lunar Reconnaissance Orbiter roughly 390,000 km (240,000 mi) away. To compensate for atmospheric interference, an error correction code algorithm similar to that used in CDs was implemented.

A two-way distance record for communication was set by the Mercury laser altimeter instrument aboard the MESSENGER spacecraft, and was able to communicate across a distance of 24 million km (15 million miles), as the craft neared Earth on a fly-by in May, 2005. The previous record had been set with a one-way detection of laser light from Earth, by the Galileo probe, of 6 million km in 1992.

Commercial Use

Various planned satellite constellations such as SpaceX's Starlink intended to provide global broadband coverage employ laser communication for inter-satellite links between the several hundred to thousand satellites effectively creating a space-based optical mesh network.

LEDs

RONJA is a free implementation of FSO using high-intensity LEDs.

In 2001, Twibright Labs released Ronja Metropolis, an open source DIY 10 Mbit/s full duplex LED FSO over 1.4 km In 2004, a Visible Light Communication Consortium was formed in Japan. This was based on work from researchers that used a white LED-based space lighting system for indoor local area network (LAN) communications. These systems present advantages over traditional UHF RF-based systems from improved isolation between systems, the size and cost of receivers/ transmitters, RF licensing laws and by combining space lighting and communication into the same system. In January 2009, a task force for visible light communication was formed by the Institute of Electrical and Electronics Engineers working group for wireless personal area network standards known as IEEE 802.15.7. A trial was announced in 2010, in St. Cloud, Minnesota.

Amateur radio operators have achieved significantly farther distances using incoherent sources of light from high-intensity LEDs. One reported 173 miles (278 km) in 2007. However, physical limitations of the equipment used limited bandwidths to about 4 kHz. The high sensitivities required of the detector to cover such distances made the internal capacitance of the photodiode used a dominant factor in the high-impedance amplifier which followed it, thus naturally forming a low-pass filter with a cut-off frequency in the 4 kHz range. Use of lasers can reach very high data rates which are comparable to fiber communications.

Projected data rates and future data rate claims vary. A low-cost white LED (GaN-phosphor) which could be used for space lighting can typically be modulated up to 20 MHz. Data rates of over 100 Mbit/s can be easily achieved using efficient modulation schemes and Siemens claimed to have achieved over 500 Mbit/s in 2010. Research published in 2009, used a similar system for traffic control of automated vehicles with LED traffic lights.

In September 2013, pureLiFi, the Edinburgh start-up working on Li-Fi, also demonstrated high speed point-to-point connectivity using any off-the-shelf LED light bulb. In previous work, high bandwidth specialist LEDs have been used to achieve the high data rates. The new system, the Li-1st, maximizes the available optical bandwidth for any LED device, thereby reducing the cost and improving the performance of deploying indoor FSO systems.

Engineering Details

Typically, best use scenarios for this technology are:

- LAN-to-LAN connections on campuses at Fast Ethernet or Gigabit Ethernet speeds.
- LAN-to-LAN connections in a city, a metropolitan area network.
- To cross a public road or other barriers which the sender and receiver do not own.
- Speedy service delivery of high-bandwidth access to optical fiber networks.
- Converged Voice-Data-Connection.
- Temporary network installation (for events or other purposes).
- Reestablish high-speed connection quickly (disaster recovery).
- As an alternative or upgrade add-on to existing wireless technologies:
    - Especially powerful in combination with auto aiming systems, this way you could power moving cars or you can power your laptop while you move or use auto-aiming nodes to create a network with other nodes.
- As a safety add-on for important fiber connections (redundancy).
- For communications between spacecraft, including elements of a satellite constellation.
- For inter-chip and intra-chip communication.

The light beam can be very narrow, which makes FSO hard to intercept, improving security. In any case, it is comparatively easy to encrypt any data traveling across the FSO connection for additional security. FSO provides vastly improved electromagnetic interference (EMI) behavior compared to using microwaves.

Technical Advantages

- Ease of deployment.
- Can be used to power devices.
- License-free long-range operation (in contrast with radio communication).
- High bit rates.
- Low bit error rates.
- Immunity to electromagnetic interference.
- Full duplex operation.
- Protocol transparency.
- Increased security when working with narrow beam(s).
- No Fresnel zone necessary.
- Reference open source implementation.

Range Limiting Factors

For terrestrial applications, the principal limiting factors are:

- Fog (10 to ~100 dB/km attenuation).
- Beam dispersion.
- Atmospheric absorption.
- Rain.
- Snow.
- Terrestrial scintillation.
- Interference from background light sources (including the sun).
- Shadowing.
- Pointing stability in wind.
- Pollution/smog.

These factors cause an attenuated receiver signal and lead to higher bit error ratio (BER). To overcome these issues, vendors found some solutions, like multi-beam or multi-path architectures, which use more than one sender and more than one receiver. Some state-of-the-art devices also have larger fade margin (extra power, reserved for rain, smog, fog). To keep an eye-safe environment, good FSO systems have a limited laser power density and support laser classes 1 or 1M. Atmospheric and fog attenuation, which are exponential in nature, limit practical range of FSO devices to several kilometres. However the free space optics, based on 1550 nm wavelength, have considerably lower optical loss than free space optics, using 830 nm wavelength, in dense fog conditions. FSO using wavelength 1550 nm system are capable of transmitting several times higher power than systems with 850 nm and are at the same time safe to the human eye (1M class). Additionally, some free space optics, such as EC SYSTEM, ensure higher connection reliability in bad weather conditions by constantly monitoring link quality to regulate laser diode transmission power with built-in automatic gain control.

VISIBLE LIGHT COMMUNICATION

Visible light communication (VLC) is a data communications variant which uses visible light between 400 and 800 THz (780–375 nm). VLC is a subset of optical wireless communications technologies.

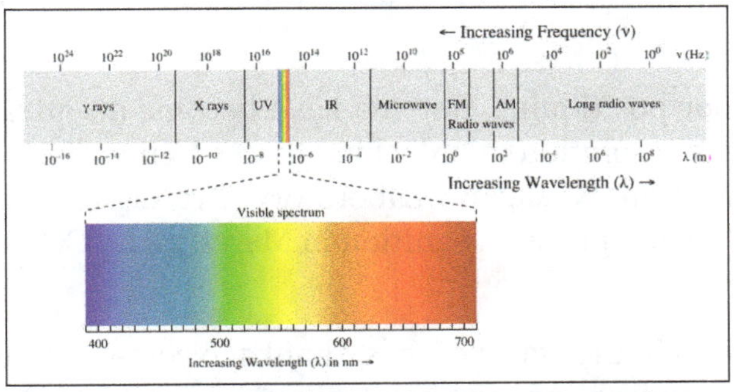

Visible light is only a small portion of the electromagnetic spectrum.

The technology uses fluorescent lamps (ordinary lamps, not special communications devices) to transmit signals at 10 kbit/s, or LEDs for up to 500 Mbit/s over short distances. Systems such as RONJA can transmit at full Ethernet speed (10 Mbit/s) over distances of 1–2 kilometres (0.6–1.2 mi).

Specially designed electronic devices generally containing a photodiode receive signals from light sources, although in some cases a cell phone camera or a digital camera will be sufficient. The image sensor used in these devices is in fact an array of photodiodes (pixels) and in some applications its use may be preferred over a single photodiode. Such a sensor may provide either multi-channel (down to 1 pixel = 1 channel) or a spatial awareness of multiple light sources.

VLC can be used as a communications medium for ubiquitous computing, because light-producing devices (such as indoor/outdoor lamps, TVs, traffic signs, commercial displays and car headlights/taillights) are used everywhere. Using visible light is also less dangerous for high-power applications because humans can perceive it and act to protect their eyes from damage.

The history of visible light communications (VLC) dates back to the 1880s in Washington, D.C. when the Scottish-born scientist Alexander Graham Bell invented the photophone, which transmitted speech on modulated sunlight over several hundred meters. This pre-dates the transmission of speech by radio.

More recent work began in 2003 at Nakagawa Laboratory, in Keio University, Japan, using LEDs to transmit data by visible light. Since then there have been numerous research activities focussed on VLC.

In 2006, researchers from CICTR at Penn State proposed a combination of power line communication (PLC) and white light LED to provide broadband access for indoor applications. This research suggested that VLC could be deployed as a perfect last-mile solution in the future.

In January 2010, a team of researchers from Siemens and Fraunhofer Institute for Telecommunications, Heinrich Hertz Institute in Berlin demonstrated transmission at 500 Mbit/s with a white LED over a distance of 5 metres (16 ft), and 100 Mbit/s over longer distance using five LEDs.

The VLC standardization process is conducted within the IEEE 802.15.7 working group.

In December 2010, St. Cloud, Minnesota, signed a contract with LVX Minnesota and became the first to commercially deploy this technology.

In July 2011, a presentation at TED Global. gave a live demonstration of high-definition video being transmitted from a standard LED lamp, and proposed the term Li-Fi to refer to a subset of VLC technology.

Recently, VLC-based indoor positioning systems have become an attractive topic. ABI research forecasts that it could be a key solution to unlocking the $5 billion "indoor location market". Publications have been coming from Nakagawa Laboratory, ByteLight filed a patent on a light positioning system using LED digital pulse recognition in March 2012. COWA at Penn State and other researchers around the world.

Another recent application is in the world of toys, thanks to cost-efficient and low-complexity implementation, which only requires one microcontroller and one LED as optical front-end.

VLCs can be used for providing security. They are especially useful in body sensor networks and personal area networks.

Recently Organic LEDs (OLED) have been used as optical transceivers to build up VLC communication links up to 10 Mbit/s.

In October 2014, Axrtek launched a commercial bidirectional RGB LED VLC system called MOMO that transmits down and up at speeds of 300 Mbit/s and with a range of 25 feet.

In May 2015, Philips collaborated with supermarket company Carrefour to deliver VLC location-based services to shoppers' smartphones in a hypermarket in Lille, France. In June 2015, two Chinese companies, Kuang-Chi and Ping An Bank, partnered to introduce a payment card that communicates information through a unique visible light. In March 2017, Philips set up the first VLC location-based services to shoppers' smartphones in Germany. The installation was presented at EuroShop in Düsseldorf. As first supermarket in Germany an Edeka supermarket in Düsseldorf-Bilk is using the system, which offers a 30 centimeter positioning accuracy can be achieved, which meets the special demands in food retail. Indoor positioning systems based on VLC can be used in places such as hospitals, eldercare homes, warehouses, and large, open offices to locate people and control indoor robotic vehicles.

There is wireless network that for data transmission uses visible light, and does not use intensity modulation of optical sources. The idea is to use vibration generator instead of optical sources for data transmission.

Colour Shift Keying

Color shift keying (CSK), outlined in IEEE 802.15.7, is an intensity modulation based modulation scheme for VLC. CSK is intensity-based, as the modulated signal takes on an instantaneous color equal to the physical sum of three (red/green/blue) LED instantaneous intensities. This modulated signal jumps instantaneously, from symbol to symbol, across different visible colors; hence, CSK can be construed as a form of frequency shifting. However, this instantaneous variation in the transmitted color is not to be humanly perceptible, because of the limited temporal sensitivity in the human vision — the "critical flicker fusion threshold" (CFF) and the "critical color fusion threshold" (CCF), both of which cannot resolve temporal changes shorter than 0.01 second. The LEDs' transmissions are, therefore, preset to time-average (over the CFF and the CCF) to a specific time-constant color. Humans can thus perceive only this preset color that seems constant over time, but cannot perceive the instantaneous color that varies rapidly in time. In other words, CSK transmission maintains a constant time-averaged luminous flux, even as its symbol sequence varies rapidly in chromaticity.

LI-FI

The Li-Fi (light fidelity) technology was proposed by the German Scientist namely Harald Haas. The main function of this technology is to transmit the data via light. This technology is idyllic for high-speed wireless communication in a restricted region, and it offers many benefits over Wi-Fi technology such as high bandwidth, ease of use, efficiency, and safety. These systems can communicate from street lights to auto-piloted cars using their headlights. As the light speed is superior hence the data communication speed is also faster in the existing system. Furthermore, this technology can be implemented for speedy data access for the laptops, and gadgets that will be transmitted during the beam in a room.

Li-Fi-light fidelity is similar to Wi-Fi technology and it is one of the future wireless communication technologies. The main feature of this technology includes fully networked, bidirectional and high-speed wireless. Nowadays, the most trending domain in wireless communication is Wi-Fi and internet users are also being increased every year. For obtaining better speed, efficiency, bandwidth, Li-Fi technology has evolved. The data transmission in this technology can be done using light because the light intensity changes quicker than the human eye for capturing. The range of data transmission in L-Fi is faster 100 times than Wi-Fi.

Li-Fi Technology.

Block Diagram of Li-Fi System

Li-Fi system mainly includes two parts namely the transmitter and receiver. The input signal at the transmitter section can be modulated with a specific time period then send the data using LED bulbs in 0's and 1's form. Here, the flashes of LED bulbs are denoted with 0's and 1's. At the receiver end, a photodiode is used to receive the LED flashes strengthens the signal & gives the output.

The block diagram of Li-Fi system is shown below, and the transmitter section includes the input, timer circuit, an LED bulb. The input of the transmitter can be any kind of data like text, voice, etc. The timer circuit in this section is used to provide the necessary time intervals among every bit, and these are transmitted to the receiver end in the form of LED flashes.

The receiver section includes photodiode as well as amplifier. Here, photodiode receives the LED bulb flashes then changes the flashes into electrical signals. Finally, the amplifier receives the signals from the photodiode and amplifies to provide the output.

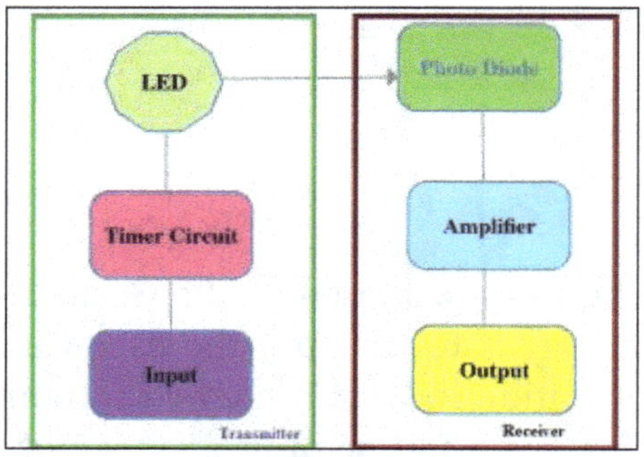

Li-Fi Block Diagram.

How Li-Fi Works?

Li-Fi is a VLC (visible light communications) system and the speed of this system is very high. Li-Fi uses normal LEDs to allow the data to transfer and increase the speed up to 224 Gigabits/sec. The data transmission of this technology can be done via illumination. The essential devices of this system are the bright light emitting diodes.

The on/off activity of LEDs permits a type of data transmission in the form of binary codes but the human eye cannot recognize this transform & the bulbs appear with a stable intensity.

Differences between Li-Fi and Wi-Fi

The differences between Li-Fi and Wi-Fi include the following:

- The Bandwidth Expansion of Wi-Fi technology is limited and exceptional for Li-Fi.
- The Speed of the Wi-Fi is150 Mbps and >10 Gbps for Li-Fi.
- The Data Density of Wi-Fi technology is low and high for Li-Fi.
- The Range of Wi-Fi is medium and low for Li-Fi.
- The Security of Wi-Fi is medium and excellent for Li-Fi.
- The power availability of Wi-Fi is low and high for Li-Fi.

- The ecological impact of Wi-Fi is medium and low for Li-Fi.
- The cost of the Wi-Fi is medium and low for Li-Fi.
- The network topologies of Li-Fi and Wi-Fi technologies is a point to point.

Advantages of Li-Fi

The advantages of Li-Fi include the following:

- Speed: The speed of the Li-Fi is very high, and we can watch the videos without buffering.
- Security: The light of the Li-Fi doesn't run through the partition, therefore, it is more protected and hacking is not possible.
- Risk-free: Li-Fi utilizes light waves which are harmless.
- Consistent: The data transfer is more protected.

Disadvantages of Li-Fi

The disadvantages of Li-Fi include the following:

- Apart from several benefits, the Li-Fi technology is facing several problems. It requires LOS (line of sight), as well as the receiver, would not be a move in inside.
- The main problem is how the receiver will send the data back to the transmitter section.
- Another disadvantage of this technology is an interference of exterior light sources such as normal bulbs; sunlight in the lane of communication will cause intermission in the transmission.
- It doesn't work in the dim areas.

Li-Fi Technology Applications

The Li-Fi router applications over Wi-Fi are promising as well as unlimited for the wireless communication which includes the following:

- Traffic Management & Road Safety.
- Medical Applications.
- Aviation.
- Communication Underwater.
- Elegant Lighting.
- Interior map-reading system for blind people.
- In dangerous Environments or In Sensitive regions.
- Transportations.
- Industrial Areas.

At present, the concept of Li-Fi technology is attracting a lot of wireless internet users due to its genuine features. Further, it can be explored for laptops; gadgets which can be transmitted through the light in the room by using Li-Fi.

References

- Communications Task Group (TG 7m) (31 May 2019). "15.7 Maintenance: Short-Range Optical Wireless". IEEE 802.15 WPANTM. Retrieved 2019-05-31
- Optical-communication, definition: techopedia.com, Retrieved 14 August, 2019
- Kachris, Christoforos; Tomkos, Ioannis (Oct 2012). "A survey on optical interconnects for data centers". IEEE Communications Surveys & Tutorials. 14 (4): 1021–1036. Doi:10.1109/SURV.2011.122111.00069
- Optical-fibre-telecommunications-basics, fibre-optics, connectivity: electronics-notes.com, Retrieved 15 January, 2019
- Yoshino, Masaki; Haruyama, Shinichiro; Nakagawa, Masao (1 January 2008). "High-accuracy positioning system using visible LED lights and image sensor". 2008 IEEE Radio and Wireless Symposium. Pp. 439–442. doi:10.1109/RWS.2008.4463523. ISBN 978-1-4244-1462-8 – via IEEE Xplore
- An-overview-of-li-fi-technology-and-its-benefits: elprocus.com, Retrieved 16 February, 2019
- "German, WWII, WW2, Lichtsprechgerät 80/80". LAUD Electronic Design AS. Archived from the original on July 24, 2011. Retrieved June 28, 2011

Applications

CHAPTER 6

Optical engineering has a wide-range of applications of optics. Its applications include optical sensors, measurement systems, lasers, fiber optic communication systems, optical disc systems, photonic devices, biomedical optics, etc. The diverse applications of optical engineering have been thoroughly discussed in this chapter.

MEDICAL APPLICATIONS OF FIBER-OPTICS

With a global population that's both growing and living longer, the world's healthcare providers are increasingly looking to advanced biomedical instrumentation to enable more efficient patient diagnosis, monitoring, and treatment. In this context, biomedical sensing applications of optical fiber are of growing importance. At the same time, recent advances in minimally invasive surgery (MIS) demand smaller disposable sensing catheters.

Endoscopic imaging applications of fiber-optics are well established, but the intrinsic physical characteristics of optical fibers also make them extremely attractive for biomedical sensing. Uncabled fibers (typically less than 250 μm diameter) can be inserted directly into hypodermic needles and catheters, so that their use can be both minimally invasive and highly localized—and fiber-optic sensors (FOS) made with them can perform remote multipoint and multiparameter sensing. Optical fibers are immune to electromagnetic interference (EMI), chemically inert, nontoxic, and intrinsically safe. Their use will not cause interference with the conventional electronics found in medical theaters. And, most importantly, the immunity of fibers to electromagnetic and radio frequency (RF) signals makes them ideal for real-time use during diagnostic imaging with MRI, CT, PET, or SPECT systems, as well as during thermal ablative treatments involving RF or microwave radiation.

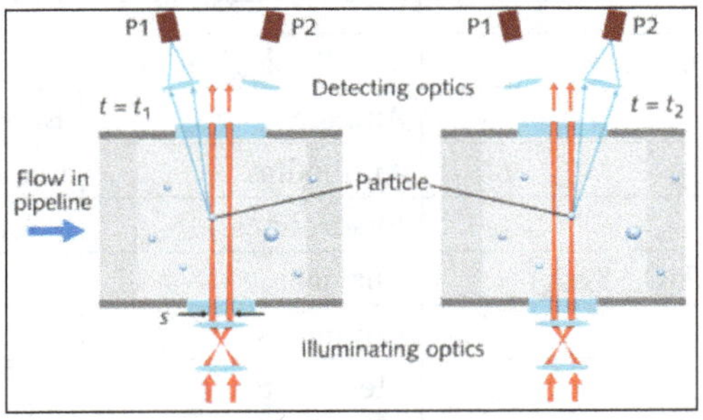

There are two basic types of optical fiber sensors. Extrinsic devices (a) relay on a transducer, whereas intrinsic devices (b) do not.

Fiber-optic Biomedical Sensors

Optical fiber sensors comprise a light source, optical fiber, external transducer, and photodetector. They sense by detecting the modulation of one or more of the properties of light that is guided inside the fiber—intensity, wavelength, or polarization, for instance. The modulation is produced in a direct and repeatable fashion by an external perturbation caused by the physical parameter to be measured. The measurand of interest is inferred from changes detected in the light property.

Fiber-optic sensors can be intrinsic or extrinsic. In an intrinsic sensor, the light never leaves the fiber and the parameter of interest affects a property of the light propagating through the fiber by acting directly on the fiber itself. In an extrinsic sensor, the perturbation acts on a transducer and the optical fiber simply transmits light to and from the sensing location.

Many different fiber-optic sensing mechanisms have been demonstrated already for industrial applications and some for biomedical applications among which are fiber Bragg gratings (FBG), Fabry-Perot cavities or external fiber Fabry-Perot interferometer (EFPI) sensors, evanescent wave, Sagnac interferometer, Mach-Zehnder interferometer, microbend, photoelastic, and others. By far the most common, however, are based on EFPIs and FBGs. Spectroscopic sensors based on light absorption and fluorescence are also common. Biomedical FOS can be categorized into four main types: physical, imaging, chemical, and biological.

Physical sensors measure a variety of physiological parameters, like body temperature, blood pressure, and muscle displacement. Imaging sensors encompass both endoscopic devices for internal observation and imaging, as well as more advanced techniques such as optical coherence tomography (OCT) and photoacoustic imaging where internal scans and visualization can be made nonintrusively. Chemical sensors rely on fluorescence, spectroscopic, and indicator techniques to identify and measure the presence of particular chemical compounds and metabolic variables (such as pH, blood oxygen, or glucose level). They detect specific chemical species for diagnostic purposes, as well as monitor the body's chemical reactions and activity. Biological sensors tend to be more complex and rely on biologic recognition reactions—such as enzyme-substrate, antigen-antibody, or ligand-receptor—to identify and quantify specific biochemical molecules of interest.

In terms of sensor development, the basic imaging sensors are the most developed. Fiber-optic sensors for measurement of physical parameters are the next most prevalent, and the least developed area in terms of successful products is sensors for biochemical sensing, even though many FOS concepts have been demonstrated.

Table: Classification of biomedical sensors by type showing various biomedical parameters of interest.

Physical	Chemical	Biological	Imaging
Body temperature	pH	Antigens	Endoscopy
Blood pressure	pO_2	Antibodies	Optical coherence tomography (OCT)
Blood flow	PCO_2	Electrolytes	Photodynamic therapy (PDT)
Heart rate	Oximetry (SaO_2, SvO_2)	Enzymes	
Force	Glucose	Inhibitors	
Position	Bile	Metabolites	
Respiration	Lipids	Proteins	
Shape sensing			

Requirements and Applications

Biomedical sensors present unique design challenges and particular problems related to their interface with a biological organism. Sensors must be safe, reliable, highly stable, biocompatible, amenable to sterilization and autoclaving, not prone to biologic rejection, and not require calibration (or at least maintain calibration for extended periods). Sensor packaging is an especially critical aspect since the devices must be very small—particularly those for implanting or indwelling purposes. The devices also must be as simple as possible.

Applications for biomedical sensors can be classified as *in vivo* or *in vitro*. *In vivo*refers to application on a whole, living organism—such as a human patient; *in vitro* refers to testing outside the body—such as laboratory blood tests. From the perspective of how sensors are applied to a patient or biological system, they can be classified as noninvasive, contacting (skin surface), minimally invasive (indwelling), or invasive (implantable). Biomedical sensors can be used in humans (clinical), in animals (veterinary), or other living organisms (life sciences), and, depending on the intended use, can be for diagnostic, therapeutic, or intensive care uses in clinical applications; research and preclinical development; or laboratory testing.

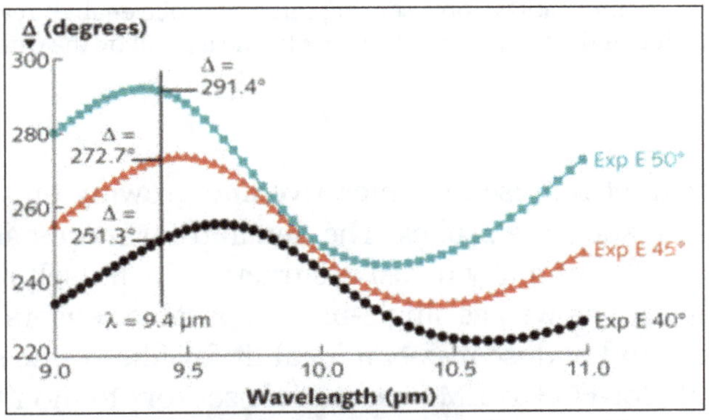

Sensors intended for implanting or indwelling applications must be very small such as this micro-miniature fiber-optic pressure sensor shown on a fingertip.

Product Developments

One of the early pioneers of fiber-optic biomedical sensors, Camino Labs, in 1984 introduced into the medical market an intracranial pressure (ICP) sensor that has since become one of the most commonly used ICP monitoring systems in the world. The device is based on an intensity modulating fiber-optic scheme relying on a miniature bellows as the transducer.

Other sensor pioneers are Luxtron with its fluoroptic temperature sensor, and FISO which has positioned itself as a leading supplier of medical fiber-optic pressure and temperature sensors. FISO's sensors are based on EFPI devices interrogated with white-light interferometry. Among a new generation of companies are Opsens, Neoptix, and Samba Sensors. By far, the most common medical FOS on the market are temperature and pressure monitors, but a handful of other diverse sensors and instruments does exist. As costs fall and new sensing techniques are developed, it's likely that the number and diversity of biomedical FOS will increase.

Among the latest development efforts are shape-sensing systems that use arrays of FBGs disposed along multicore, singlemode fibers. The FBGs will shift peak wavelengths in response to the strain and curvature stress produced during bending. The fiber arrays help determine the precise position and shape of medical tools and robotic arms used during MIS. Companies pursuing such developments are Hansen Medical, Intuitive Surgical, Luna Innovations, Measurand, and Technobis.

Another relevant new FOS product in precertification trials is the EndoSense TactiCath force-sensing catheter. Fiber Bragg grating sensors are mounted on the tip of an intra-aortic catheter that also serves as a laser-ablation delivery probe for the treatment of atrial fibrillation. The FBGs detect the force exerted against the heart wall by the stress induced on them. Force control is essential for delivering appropriate laser ablation pulses needed to produce lesions that are induced in the heart.

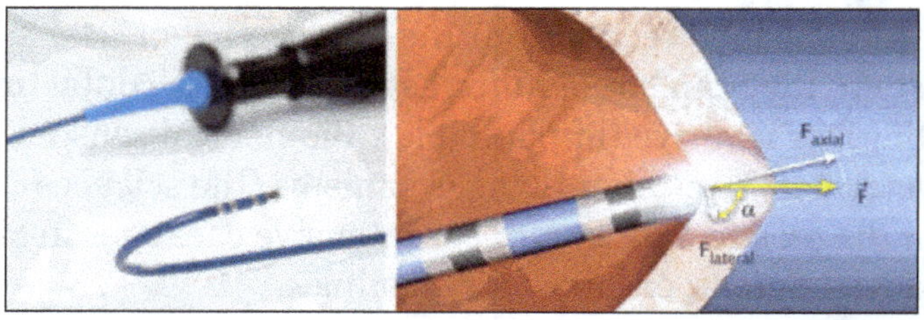

A fiber-optic intra-aortic force sensing catheter probe enables real-time monitoring of the force exerted against the heart wall by the catheter.

Outlook

The biomedical sensing market represents a lucrative and growing opportunity for FOS, particularly for large volumes of disposable probes. The demand for more patient monitoring devices combines with a trend toward minimally invasive surgery, which itself requires a variety of minimally invasive medical devices as well as single-use, disposable sensors of small size that can be incorporated into catheters and endoscopes—an ideal fit for fiber-optic sensors. There is also an unquestionable opportunity for FOS as EMI-compatible sensors to monitor vital signs during use of MRI (and related techniques), as well as RF treatments.

BCC Research (Wellesley, MA) estimates the US market for patient monitoring devices to be worth \$3.6 billion in 2007, and to reach \$5.1 billion in 2013. The disposable sensors (and other consumables) portion of the market was estimated to be \$2.6 billion in 2007 and to reach ~\$3.4 billion in 2013. The FOS share of this global market is small and estimated to be at around \$100 million. However, the potential is tremendous and fiber-optic biomedical sensors provide capabilities and features that cannot be otherwise obtained. High cost remains a barrier, however, as does the lengthy development cycles and required regulatory process. Sensor design and development is not trivial, and proper material selection, design, biocompatibility, patient safety, and other issues must be taken into account. Nevertheless, there are already several successful products in the market and more to come.

OPTICAL METROLOGY

Optical metrology is the science and technology concerning measurements with light. Such measurements can either target properties of light and light sources or properties of objects such as dimensions, distances and temperatures. There is no strict boundary between those fields, because often one uses measured properties of light not just to characterize a light source, but for other purposes – for example, optical frequency metrology is used for ultraprecise optical clocks.

Some examples of optical metrology are:

- Optical distance measurements with lasers can be based on, e.g., interferometers or measurements of the time-of-flight of light pulses. This is an example for dimensional metrology.
- Optical profilometers are widely used for measuring surface topographies, e.g. in semiconductor chip production and for the quality control in optical fabrication. Form metrology also uses various other kinds of instruments for measuring surface shape (contour) and surface roughness.
- Optical time-domain reflectometers are used for inspecting fiber-optic links – for example, finding faulty fiber splices or fiber connectors.
- Optical powers can be measured with photodiodes, thermal power meters, or other equipment. Optical irradiance and other illumination measurements can address either some pure physical quantity such as an optical intensity (power per unit area) (radiometry) or something like a perceived brightness (photometry). Integrating spheres are utilized for radiation emitted in a wide range of direction.
- Spectral optical properties are measured with devices like spectrographs or other spectrometers, wavemeters and self-heterodyne setups.
- Optical frequency metrology deals with high-precision measurements of optical frequencies. One can produce ultraprecise optical clocks, surpassing the performance of cesium atomic clocks.
- Optical temperature sensors may be based on the analysis of the thermal emission of hot bodies, or rely on the measurement of occupation probabilities for energy levels of atoms or molecules.
- Fiber-optic temperature and strain sensors allow for distributed sensing, often of temperature and strain combined. They can be used, for example, for measurements in industrial processing plants, bridges and tunnels, buildings, oil and gas pipelines and power transmission lines.

Typical Qualities of Optical Metrology

In many cases, optical metrology can be extremely precise and is ultimately limited by laser noise or quantum noise in detection.

Optical measurements are usually quite fast and suitable e.g. for in-process metrology, i.e., for monitoring industrial production processes.

Generally, optical measurements are non-destructive. Even very sensitive parts can be checked without touching them (non-contact methods), i.e., without a risk of damage.

AUGMENTED REALITY AND VIRTUAL REALITY

The ultimate augmented reality (AR) display can be conceived as a transparent interface between the user and the environment—a personal and mobile window that fully integrates real and virtual information such that the virtual world is spatially superimposed on the real world. An AR display tailors light by optical means to present a user with visual (symbolic or pictorial) information superimposed on spaces, buildings, objects, and people.

These displays are powerful and promising because the augmentation of the real world by visual information can take on so many forms. For example, augmentation can add identifying and highlighting features on moving targets in the environment, zoom in on critical information to aid decision-making, or provide hidden information not visible to the human eye, but instead available from text, data, or model databases.

Data may be acquired via a wide range of technology, mimicking super-human characteristics, such as polarization sensing or infrared and terahertz imaging. By allowing us to paint anything we know on the spaces, objects, and people in the world, these displays extend our capability to see beyond our natural senses and bring any information to bear on a user's context and actions.

Unfortunately, optical and optoelectronic engineers differ in their approaches to developing the ultimate AR display that is physically comfortable, prevents eye fatigue, presents the desired augmented information without obscuring reality, and does it all at a price point bearable to the average consumer or to the industrial user as dictated by the application. Fortunately, designers are meeting these challenges in step with new optics and photonics advances that are enabling AR displays to mature and thrive.

Applications

Consider several emerging applications of AR. Boeing (among others) has been building a better future for their technicians—setting up a roadmap since the 1990s—by providing them smart AR glasses while installing electrical wiring in aircraft. Superimposing text of the internal schematics of the plane on their AR displays during repair and maintenance boosts both worker productivity and the likelihood of error-free work.

While the right to privacy of information is making news headlines, consider superimposition of personal information made voluntarily public in the cloud. As you enter a space, your display interface may communicate high-level visual information about the people around you. This information could potentially facilitate stimulating personal discussions and interactions.

Another compelling example of augmentation is in the medical domain. Practitioners can see the superimposition of data or models from databases revealing the internal anatomy of patients acquired via medical devices. Using tightly superimposed spatial information about the human tissues of interest, surgeons of the future have the potential to reduce the length of surgeries and critically improve outcomes. By guiding the scalpel with data, the augmentation may lower the risk of malpractice associated with cutting unwanted structures. Consider brain surgery, where a small mistake may permanently damage the facial nerves—a tragic spatial error that AR displays could one day eradicate.

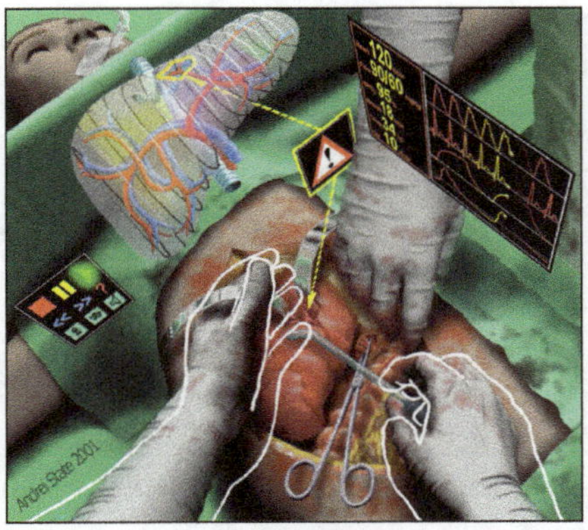

Augmented reality can play a particularly useful role in surgery by overlaying critical information about the patient during the actual surgery itself to improve patient outcomes.

The Road to the Ultimate Display

It was back in the 1960s that Ivan Sutherland conceived and prototyped what he called the ultimate display—the first AR stereoscopic head-worn display (HWD)that was tethered and mechanically tracked. The next generation of developments focused on military helmet-mounted displays (HMDs), whose increasingly higher levels of sophistication continue to the present. In parallel, new applications called for the birth of fully mobile HWDs that were first targeted in the 1990s for maintenance and medical augmentation and continue to develop and advance to this date.

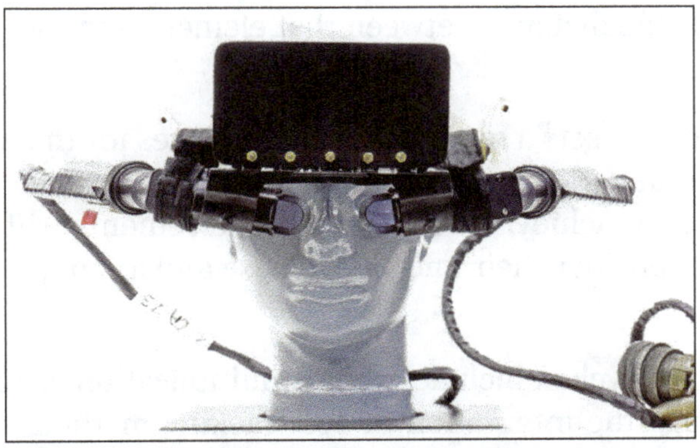

Sutherland's AR 1960s Ultimate Display, which is presumed to be the first head-mounted stereoscopic display; it was called the Sword of Damocles because it was suspended from the ceiling to reduce the weight pressing on the wearer's head.

Now, AR is coming to the individual user spurred by the billion-dollar market potential of applications supporting a wide range of consumers. The race is on to create a robust, untethered, and aesthetical AR display that acts as a window into the world. The ultimate goal is to make that mobile window larger by achieving a wide field of view (FOV)—60° or larger—with full color and perfectly embedded in a trendy eyeglass or sunglass format.

The glasses must account for the tendency of one's eyes to naturally scan behind the optics within the boundary of the eyebox—that near-eye volume in space within which our eyes may scan freely without losing track of the digital world. Without a sufficiently large eyebox, the experience of losing track of augmented information may feel to the user like a blackout (a sudden erasure of the virtual information)—a fault that could inhibit adoption of the technology.

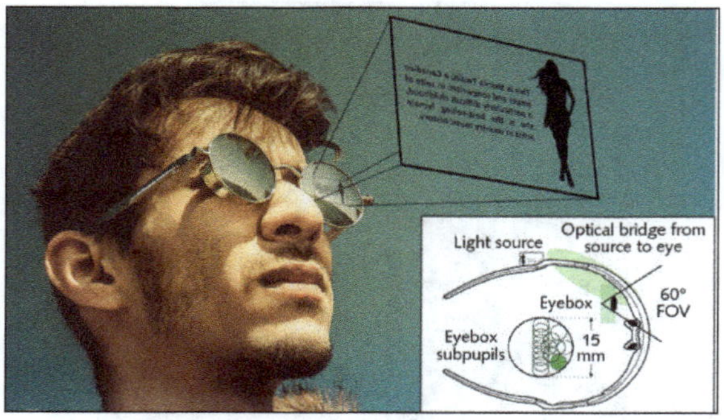

The 21st Century Ultimate AR Display is shown.

Challenges for Optical Designers

Optical designers of wearable displays understand the daunting challenge of meeting the demanding specifications of a large FOV together with a large eyebox, while at the same time satisfying the constraints for robustness, packaging, low cost, and the high-yield manufacturing processes that are essential for the viability of a consumer product. In addition, the foregoing features must be power-efficient to maximize the use time from the AR display battery.

The optical concept of the Lagrange invariant (or etendue, in the case of the generalization to three dimensions) establishes that the mathematical product of the FOV angle and eyebox size will dictate the complexity of the optical system. Furthermore, the size of the first optical element with respect to the eyes is set by the distance between that element and the eye multiplied by the FOV angle.

The eyeglass or sunglass format sets a range of acceptable sizes for that component. Then, a range of FOVs and distances can help determine a range of solutions. If the FOV is non-negotiable in terms of achieving ≥60°, then achieving the eyebox requirements will be the challenge. Optical designers should carefully consider their knowledge base and assumptions before setting off on a particular design path.

Approaches to expanding the eyebox include eyebox multiplication or replication techniques. The designer might also leverage the integration of eye-tracking methods to dynamically adjust the location of the display's eyebox in real time to the current position of the user's iris as the eyes gaze around the merged virtual and physical world. The latter approach must be accurate and robust to accommodate a wide diversity of users. These approaches to creating an effective larger eyebox can add significant complexity to the engineering challenge of seamlessly tailoring light in an AR display.

The Intersection of AR Displays and Human Factors

The ultimate AR display is part of the user's body. The interface is a window to interacting not only with objects and things, but also with other people. While knowledge of human factors is central to creating a visual experience that is as natural and intuitive as possible, a wearable display will only gain social acceptance if it encompasses both aesthetics and usability to support social interaction. In 2016, R. E. Patterson and I discussed the intention behind the use of information displays as it relates to cognitive engineering, which involves the meaning of display content.

A ubiquitous AR system design for cognitive augmentation arouses considerations of cognitive and social-technical issues.

Questions that Guide Display Development

Provided an optical specification where certain parameters and performance metrics are seen as a target, other specifications may be set as a goal. For example, what type of light source will be adopted? Will an extended microdisplay (with or without illumination optics) be used with custom optics to image in one shot the full-color image onto the retina, or will a point-like light source coupled to a scanning device such as a microelectromechanical systems (MEMS)-based scanner or a scanning fiber tip be used to sequentially draw point by point the image at high speed on the retina in multiple colors? For the auxiliary optics, will thin waveguides be stacked to confine light in a thin sheet in front of the eyes, and if so, are there approaches to curved waveguides that improve the aesthetic form factor? How will the light from each color be coupled into the waveguide and output to the eyes of the user, and will the system suffer from crosstalk or stray light that may add unacceptable fog to an otherwise-sharp image across the full FOV? Can the waveguide be manufactured at high yield, as it has been a common issue with various approaches adopted to this date? How authentic will the view of the real world be through the waveguide? Or will the approach leverage the sunglass interface with a conventional or holographic-like combiner as an optical element that combines the real and digital world?

Ultimately, the answers to these questions sketch a view of the holy grail of the ultimate AR display. A challenge is to establish a seamless geometry of all optical components involved to yield the desirable image quality within the targeted 60° FOV in an eyeglass or sunglass format that a large pool of users will absolutely want to wear.

Enter Freeform Optics

This is where the emerging technology of freeform optics is expected to revolutionize the AR solution space. The phrase 'freeform surface' has been used in illumination optics for some time, where complex, continuous, or segmented surfaces have been designed to tailor light. In the 1980s, Polaroid Corporation was the first to adopt a freeform prism in the viewfinder of its SX70 Single Lens Reflex camera. The freeform prism was then further developed in the 1990s for virtual reality (VR) headsets, but unfortunately these prisms were unable to create wide FOVs for immersion.

In AR for the consumer market, however, one must truly assess whether looking at the real world through a thick piece of plastic (a low-cost and lightweight molded prism) will ever be conducive to consumer adoption. It is also worth noting that the most common thinner geometry is in fact not compliant with optical aberration correction over moderate to large FOVs. This issue was recently discovered when applying advanced lens design methods for freeform optics. At a time when users seek high-definition image quality with a large FOV in an aesthetic display format, a different pathway is sought.

Freeform optics expands well beyond the freeform prism since freeform surfaces may be combined in multiple geometries, providing the ability to fold optical systems in three dimensions while simultaneously providing for a finite range of viable folding geometries together with the ability to correct for both blurring and warping optical aberrations. On a different scale when engineering complex surfaces, sub-wavelength periodic nanostructures provide another pathway to creating optical components that have been shown to provide unique means to shape the light field.

While warping aberrations may be mitigated in software, iris location behind the optics may cause space distortions if not accounted for. Naturally, eye-movement recording within an AR display tracks the convergence of the eyes to provide in-focus data at the gaze point together with appropriately blurred contextual information that more naturally combines with real-world vision. Beyond this function, eye tracking can also extract more information about the user, such as understanding what a user looks at while providing a means to measure other user states such as awareness and fatigue.

The ultimate AR display seeks to be seamless, always accompanying the user, merging virtual and physical spaces, and becoming as common, necessary, and ubiquitous as smartphones. As we learn more about human vision and cognitive processes associated with a wide range of applications, it is certain that optical technologies will continue to emerge and combine in unique ways to seamlessly guide the light while widening the window of the ultimate AR display of the future.

APPLICATION OF ADVANCED FIBER-OPTIC COMPONENTS IN AEROSPACE AND DEFENSE

Recent advances in fiber-optic technology have led to a growing demand for optical data links and sensors in aerospace and defense applications. In comparison with copper-based data links, which have been deployed extensively in military applications, fiber-optic data links offer a number of advantages.

The design of new military and aerospace vehicles imposes many restrictions on size and weight of components, while also requiring many more embedded sensors for monitoring the condition of the airframe and critical systems. The routing of these systems is significantly complicated by the available space and weight allowance inside the vehicle structure.

New materials have also affected the way modern vehicles are manufactured. Advanced composite materials are becoming commonplace in modern vehicles. While these materials can save a lot of weight, they are effectively transparent to electromagnetic energy like radio waves. These vehicles can therefore be more susceptible to electromagnetic interference than the more traditional metal structures that acted in part like a Faraday cage to partially protect the electronics inside.

Shipboard communication networks have also evolved dramatically in recent years. The communication infrastructure found on many ships is now more comparable to what previously might have serviced an entire city. While legacy systems might have run at 155 Mbps, new shipboard technologies can pack 10 or more channels on a single optical fiber, each channel supporting 10 Gbps. Fiber-based optical networks offer a number of significant advantages for defense and aerospace applications over conventional copper-based networks. It has been demonstrated that optical links have the capability to transport mixed digital and analog signals at speeds in excess of 10 Gbps, compared to the 100 Mbps that is more typical of copper connections. Along with higher data rates, optical fibers can also transmit light with losses as low as 0.2 dB/km, which is effectively lossless for most ship and aircraft applications, allowing power budgets to be kept low. Traditional fiber-optic components have some limitations in aerospace and defense applications.

Strengths and Limitations of Fiber-optics

Fiber-optics allow the use of Wavelength Division Multiplexing (WDM) technology, where many different channels can be transmitted down the same fiber by sending them on different colors (or wavelengths) of light. This can significantly reduce the size, weight, and complexity of cable routing.

Since only light signals are traveling along these optical fibers, they are effectively immune to electromagnetic interference. Moreover, optical fiber presents no spark or fire hazard, and can even be routed alongside or through fuel bays if required.

One of the most significant drawbacks of fiber-optics has been the reliance on traditional micro-optic assemblies used in most applications to date. Virtually all optical components to date have been built using miniature lenses, filters, and other optical components that are aligned by hand and glued in place. There are several potential issues with this approach.

First, the labor involved with aligning these micro-optic components is a very significant portion of the cost, which has resulted in nearly all this assembly being done overseas. Also, many piece parts epoxied together presents a risk of a subcomponent being knocked out of alignment in harsh environments, high vibration, or mechanical shock. This is particularly important in aerospace applications where vibration or shock up to 20 *g* acceleration is not uncommon. Finally, the amount of functionality that can be included in a single optical module is significantly limited by the sheer number of subcomponents that can be reliably aligned in a single package.

Planar Lightwave Circuit or *PLC* optical-chip technology addresses all of these shortcomings. This PLC platform collapses many of the same optical functions onto a silicon chip. Custom optical circuits can be designed and developed for virtually any application. These optical chips look visually similar to their electronic chip counterparts, as shown in figure, and are in fact fabricated using very similar processes. However, along with the typical electrical connections found on any silicon chip, these PLC chips also have optical connections that receive or transmit signals optically into and out of the PLC chip.

Similar to their electronic counterparts, optical chips are fabricated on silicon wafers, making them small, lightweight, and rugged.

This optical chip is housed in a small package designed for soldering onto a common printed circuit board. The result is a very compact, lightweight optical module that is effectively a single chip when assembled, minimizing any chance of parts becoming misaligned during high vibration or shock. A complete bidirectional transceiver capable of transmitting and receiving data at 2.5 Gbps can be integrated on a chip that is smaller than 4 x 10 mm, weighing less than 5 grams. These chip-based transceivers can withstand harsh g-forces, including sinusoidal vibrations with amplitude of 20 *g* acceleration, followed by mechanical shock of 500 *g* acceleration. The components are capable of temperature cycling through a range of more than 125 °C.

PLCs Deliver Functionality not Possible in other Platforms

While virtually any optical function can be integrated on the silicon chip, in some cases a hybrid approach is advantageous, where small chips are also integrated onto the PLC platform. For example, very high-performance lasers can be fabricated in indium phosphide and integrated onto the PLC as part of a hybrid approach. This allows the optical chip to provide state-of-the-art performance, leveraging different material systems as needed.

All of this hybrid integration can be accomplished using robotic pick-and-place machines typically used for electronic assembly. These customized machines can bond a new chip every 20 seconds, with extremely good repeatability, making them ideally suited for volume production. Equally important, this approach requires very minimal hands-on labor, allowing these products to be manufactured domestically.

The photonic integration technology allows the development of ultra compact and lightweight communication components such as optical transceivers. Optical transceivers are essential components in fiber-optic communication systems. They contain both a transmitter and a receiver in a single housing, and can carry upstream and downstream mixed digital and analog signals through a single fiber.

At present, the optical transceivers designed and produced for aerospace and defense applications are built using traditional micro-optic technology. The complexity of assembling a large number of critically aligned micro-optic components in a single hermetic package makes these transceivers expensive and susceptible to failure in harsh military applications.

Planar Lightwave Circuit technologies have the capability to deliver outstanding improvements in the reliability and cost of optical transceivers, and represent a major step forward in aerospace and defense communication systems.

Early Applications in Aerospace and Defense

As an example, this rugged PLC platform is ideally suited for the development of multichannel redundant transceivers. Photonic technology originally developed for the telecommunication industry has been leveraged to combine multiple signals at different wavelengths onto a common fiber, using proprietary filtering technology that is integrated into the PLC chip, requiring no external filters or lenses. If one transceiver fails, the others still provide support for critical avionics systems. The transceiver shown in figure is capable of 2.5 Gbps transmission, while simultaneously receiving downstream traffic at 2.5 Gbps, while also receiving analog video signals on a third channel.

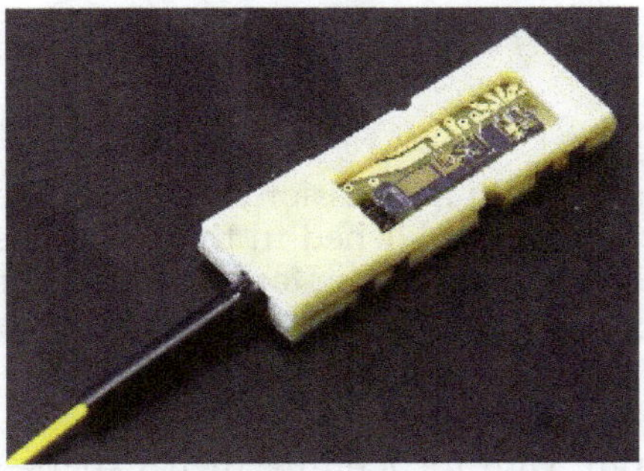

The compact, lightweight, bidirectional optical transceiver can transmit and receive simultaneously at 2.5 Gbps, while also receiving analog video signals at up to 1 GHz on a third channel.

This transceiver is packaged in a very compact housing, and designed for easy mounting and soldering onto a standard printed circuit board. This transceiver module was originally developed for use in Unmanned Aerial Vehicles (UAVs), although it has since been customized for a number of other applications.

While the transceiver shown in figure supports three channels each running at up to 2.5 Gbps, some applications require even more capacity. The very flexible PLC platform is fully scalable in both channel count and data rate. Enablence Technologies manufactures PLC-based multichannel receiver modules that include an 8-channel wavelength division multiplexer, in addition to 8 receiver channels, each running over 10 Gbps, all integrated on a silicon platform. This receiver module, along with a matching transmitter module, is designed to form the basis of a shipboard communications network. Even more of the manufacturing efforts at Enablence Technologies are focused on custom modules unique to specific applications.

Some basic examples of optical modules that can be developed based on Planar Lightwave Circuits have been examined herein. It is not uncommon for these modules to be significantly more advanced, often containing dozens or even hundreds of optical elements on a single chip. This level of functionality is typically not even feasible using more traditional optical platforms, and is opening up new applications for fiber-optics in aerospace and defense applications, including communications, sensors, and monitoring. When reviewing some of the strengths and limitations of traditional fiber-optics, it becomes clear how PLCs are poised to displace these technologies in present and future defense applications.

References

- Medical-applications-of-fiberoptics-optical-fiber-sees-growth-as-medical-sensors, fiber-optics: laserfocusworld.com, Retrieved 17 March, 2019
- Optical-metrology, rp-photonics.com, Retrieved 19 April, 2019
- Arvr-displays-engineering-the-ultimate-augmented-reality-display-paths-towards-a-digital-window-into-the-world, optics: laserfocusworld.com, Retrieved 20 May, 2019
- Advanced-enable-new-applications-aerospace-defense: mil-embedded.com, Retrieved 21 June, 2019

PERMISSIONS

All chapters in this book are published with permission under the Creative Commons Attribution Share Alike License or equivalent. Every chapter published in this book has been scrutinized by our experts. Their significance has been extensively debated. The topics covered herein carry significant information for a comprehensive understanding. They may even be implemented as practical applications or may be referred to as a beginning point for further studies.

We would like to thank the editorial team for lending their expertise to make the book truly unique. They have played a crucial role in the development of this book. Without their invaluable contributions this book wouldn't have been possible. They have made vital efforts to compile up to date information on the varied aspects of this subject to make this book a valuable addition to the collection of many professionals and students.

This book was conceptualized with the vision of imparting up-to-date and integrated information in this field. To ensure the same, a matchless editorial board was set up. Every individual on the board went through rigorous rounds of assessment to prove their worth. After which they invested a large part of their time researching and compiling the most relevant data for our readers.

The editorial board has been involved in producing this book since its inception. They have spent rigorous hours researching and exploring the diverse topics which have resulted in the successful publishing of this book. They have passed on their knowledge of decades through this book. To expedite this challenging task, the publisher supported the team at every step. A small team of assistant editors was also appointed to further simplify the editing procedure and attain best results for the readers.

Apart from the editorial board, the designing team has also invested a significant amount of their time in understanding the subject and creating the most relevant covers. They scrutinized every image to scout for the most suitable representation of the subject and create an appropriate cover for the book.

The publishing team has been an ardent support to the editorial, designing and production team. Their endless efforts to recruit the best for this project, has resulted in the accomplishment of this book. They are a veteran in the field of academics and their pool of knowledge is as vast as their experience in printing. Their expertise and guidance has proved useful at every step. Their uncompromising quality standards have made this book an exceptional effort. Their encouragement from time to time has been an inspiration for everyone.

The publisher and the editorial board hope that this book will prove to be a valuable piece of knowledge for students, practitioners and scholars across the globe.

INDEX